SpringerBriefs in Computer Science

SpringerBriefs present concise summaries of cutting-edge research and practical applications across a wide spectrum of fields. Featuring compact volumes of 50 to 125 pages, the series covers a range of content from professional to academic. Typical topics might include:

- A timely report of state-of-the art analytical techniques
- A bridge between new research results, as published in journal articles, and a contextual literature review
- A snapshot of a hot or emerging topic
- An in-depth case study or clinical example
- A presentation of core concepts that students must understand in order to make independent contributions

Briefs allow authors to present their ideas and readers to absorb them with minimal time investment. Briefs will be published as part of Springer's eBook collection, with millions of users worldwide. In addition, Briefs will be available for individual print and electronic purchase. Briefs are characterized by fast, global electronic dissemination, standard publishing contracts, easy-to-use manuscript preparation and formatting guidelines, and expedited production schedules. We aim for publication 8–12 weeks after acceptance. Both solicited and unsolicited manuscripts are considered for publication in this series.

More information about this series at http://www.springer.com/series/10028

Azad Naik • Huzefa Rangwala

Large Scale Hierarchical Classification: State of the Art

Azad Naik
Microsoft (United States)
Redmond, WA, USA

Huzefa Rangwala
George Mason University
Fairfax, VA, USA

ISSN 2191-5768 ISSN 2191-5776 (electronic)
SpringerBriefs in Computer Science
ISBN 978-3-030-01619-7 ISBN 978-3-030-01620-3 (eBook)
https://doi.org/10.1007/978-3-030-01620-3

Library of Congress Control Number: 2018957841

This Springer imprint is published by the registered company Springer Nature Switzerland AG
The registered company address is: Gewerbestrasse 11, 6330 Cham, Switzerland

This book is dedicated to our lovely family members and to all data mining, machine learning and artificial intelligence learners.

Preface

This book evolved from the several years of research done by authors in the field of hierarchical classification. It provides a comprehensive overview of the recent advancements in large scale hierarchical classification (LSHC) along with state-of-the-art techniques. LSHC has gained significant interest amongst researchers and academia due to its profound applications in various fields for organizing large stream datasets aka Big Data. Dealing with LSHC can be challenging due to several issues that arise in the large scale settings. This book aims to provide understanding of the issues along with solutions to solve it. In particular, more focus has been on two of the most compressing HC problem—hierarchical inconsistency and feature selection.

This book is intended to benefit the readers with intermediate expertise in data mining having a background in classification (supervised learning). This book would be helpful for students and researchers working in multiple disciplines including computer science and engineering, computational science, bioinformatics, statistics, etc. We hope you will enjoy reading it.

Redmond, WA, USA
Fairfax, VA, USA
August 2018

Azad Naik
Huzefa Rangwala

Acknowledgements

Writing a book is harder than we thought and more rewarding than we could have ever imagined. None of this would have been possible without the love, encouragement and support from our beloved family members.

We would like to express our sincere thanks to Richa Parihar who has provided immense help with preparation of figures. Her assistance has been invaluable.

We would also like to thank Springer for the excellent support during the different stages of preparation of this book, and we would like to thank the senior editor, Susan Evans, for her support and professionalism.

Contents

Acronyms

μF1	Micro-F1
BLF	Bottom Level Flattening
CATH	Class, Architecture, Topology and Homologous
DMOZ	Directory Mozilla
FC	Flat Classification
FS	Feature Selection
GC	Global Classification
HBLR	Hierarchical Bayesian Logistic Regression
HC	Hierarchical Classification
HD	High Distribution
hF1	Hierarchical F1
HOT	Hierarchical Orthogonal Transfer
hP	Hierarchical Precision
hR	Hierarchical Recall
HSVM	Hierarchical Support Vector Machine
INF	Inconsistent Nodes Flattened
IPC	International Patent Classification
*k*NN	*k*-Nearest Neighbor
LCL	Local Classifier per Level
LCN	Local Classifier per Node
LCPN	Local Classifier per Parent Node
LD	Low Distribution
LR	Logistic Regression
LSHC	Large Scale Hierarchical Classification
LSHTC	Large Scale Hierarchical Text Classification
MF1	Macro-F1
MHMTL	Multiple Hierarchy Multi-task Learning
MLF	Multiple Level Flattening
MTL	Multi-task Learning
SCOP	Structural Classification of Proteins
SHMTL	Single Hierarchy Multi-task Learning

SSL	Semi-Supervised Learning
STL	Single Task learning
SVM	Support Vector Machine
TD	Top-Down
TE	Tree-Induced Error
TL	Transfer Learning
TLF	Top Level Flattening

Chapter 1
Introduction

Data is everywhere, and it's generated in various forms (such as texts, images, videos) and through various means (such as social network, digital devices, Internet, sensors). The amount of data available is massive. It's growing continuously year after year at an exponential rate in almost every field ranging from astronomical data to biological and web data. In fact over the last 2 years alone, 90% of the data in the world was generated. According to Domo's data never sleeps report,[1] numbers are staggering for data generated per minute by users across different services. For example, 456,000 tweets are tweeted on Twitter, 154,200 Skype calls are made, 527,760 photos are shared on Snapchat, and 4,146,600 YouTube videos are watched every minute.

In order to extract useful and meaningful information from data, we need to first organize and structure them. Hierarchical structures/taxonomies provide a natural and convenient way to organize information. Data organization using hierarchy has been extensively used in several real-world application domains such as gene taxonomy[2] for organizing gene sequences, DMOZ taxonomy[3] for structuring web pages, international patent classification hierarchy[4] for browsing patent documents, audio taxonomy for organizing music signals [3], and ImageNet[5] for indexing millions of images.

Definition 1.1 Hierarchical structure (or hierarchy) is defined as a set of objects V and a partial order $\prec$ over the pairs of elements of V. Partial ordering comes from

[1] https://www.domo.com/learn/data-never-sleeps-5?aid=ogsm072517_1&sf100871281=1.

[2] http://geneontology.org/.

[3] http://www.dmoz.org/.

[4] http://www.wipo.int/classifications/ipc/en/.

[5] http://www.image-net.org/.

A. Naik, H. Rangwala, *Large Scale Hierarchical Classification: State of the Art*, SpringerBriefs in Computer Science, https://doi.org/10.1007/978-3-030-01620-3_1

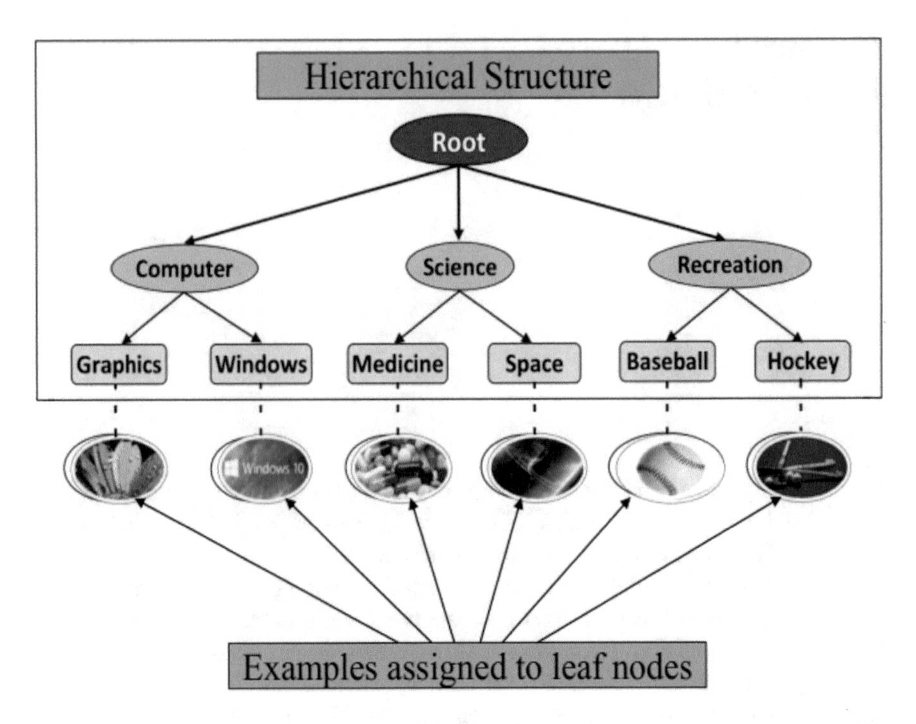

Fig. 1.1 Hierarchical structure containing two levels (excluding root node), three internal nodes, and six leaf nodes. Nodes at higher levels contain generic classes, and it becomes more and more specific as we go down the level. Examples are assigned to leaf nodes in the hierarchy (shown by dotted lines)

the parent-child relationship between the elements of V. The following properties hold true for the hierarchy:

1. Asymmetry—If $v_i \prec v_j$, then $v_j \nprec v_i, \forall v_i, v_j \in V$.
2. Anti-reflexivity—$v_i \nprec v_i, \forall v_i \in V$.
3. Transitivity—If $v_i \prec v_j$ and $v_j \prec v_k$, then $v_i \prec v_k, \forall v_i, v_j, v_k \in V$.

> Hierarchical structure assumes that each node in the hierarchy is a generic type of its children nodes and specific type of its parent node, thereby implying a hierarchical relationship.

Figure 1.1 shows an example of small hierarchical structure with two levels of hierarchy (excluding root node), three internal nodes, and six leaf nodes with instances (or examples) assigned to them. Two main type of relationships exist within the hierarchy:

1. Parent-child relationship: Two neighboring nodes directly connected by an edge have parent-child relationship between them. For example, in

Fig. 1.1: *Root—Science*, *Computer—Graphics*, *Recreation—Baseball* have parent-child relationship.
2. Siblings relationship: Nodes with common parents have siblings relationship between them. For example, in Fig. 1.1: *Computer—Science—Recreation*, *Windows—Graphics*, *Medicine—Space* have siblings relationship.

Definition 1.2 Root node—Node in the hierarchy with no parent and at least one child is known as root node.

Complex hierarchies can have more than one root node. For simplicity, this book considers hierarchical structure having one root node only.

Definition 1.3 Internal node—Node in the hierarchy having at least one parent and at least one children is known as internal node. Internal node is also referred as non-terminal node. In Fig. 1.1—*Computer*, *Science* and *Recreation* node belongs to internal node.

Definition 1.4 Leaf node—Node in the hierarchy having at least one parent and no children is known as leaf node. In other words, node that doesn't belongs to root or internal node is leaf node. Leaf node is also referred as terminal node. In Fig. 1.1—*Graphics*, *Windows*, *Medicine*, *Space*, *Baseball* and *Hockey* node belongs to leaf node.

Once hierarchical structure is available for the dataset, next step is to develop classifiers (or models) that can automatically classify newly generated unlabeled examples (or instances) into different nodes within the hierarchy, accurately and efficiently. Developing classifiers that involves hierarchy with thousands of classes is not an easy task due to several challenges discussed in Sect. 1.2.

1.1 Large Scale Hierarchical Classification Problem

Classifying unlabeled instances manually is a nontrivial task. In manual classification, users (typically experts in the field) interpret the meaning of unlabeled instances, identify the relationships between concepts, and do categorization. While this gives users more control over classification, manual classification is both expensive and time-consuming. Moreover, it's not feasible for large-scale datasets. To overcome this, automated classification that applies machine learning techniques for classification is desired. This results in faster, scalable solution for handling large volumes of data. Figure 1.2 depicts the difference between manual and automated classification.

Given, a hierarchy containing thousands of classes (or categories) and millions of instances (or examples), there is an essential need to develop an efficient and automated machine learning approaches to categorize unlabeled (or unknown)

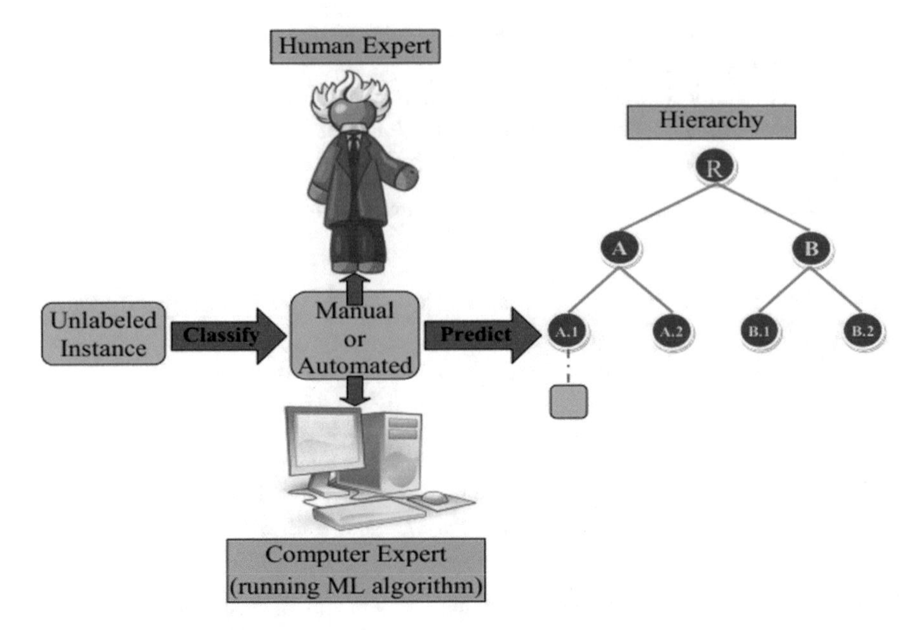

Fig. 1.2 Difference between manual and automated classification. In manual classification, humans are experts, whereas in automated classification computers (running machine learning (ML) algorithms) are experts

instances into hierarchy of classes. This problem is referred to as large-scale hierarchical classification (LSHC) task. It is an important machine learning problem that has been researched and explored extensively in the past few years [1, 2, 11, 12]. The objective of this book is to provide a comprehensive overview of the various developed approaches for dealing with LSHC problem.

1.2 Challenges with Large Scale Hierarchical Classification

Manual annotation of unlabeled instances into hierarchy of classes is a tedious and cumbersome task. This problem become even more difficult with the exponential growth rate of data over time. Although, several traditional binary and multi-class classification techniques have been developed for automated classification, they are not effective (and scalable) for LSHC problems because they ignore the implicit inter-class relationships information that are available from the hierarchical structures. To overcome this shortcoming, various HC approaches have been proposed in the literature [1, 2, 4, 5, 7, 11, 13]. Although HC approaches improve performance, there are several factors that make LSHC challenging. In the rest of the chapter, we will discuss and highlight some of the challenges associated with LSHC.

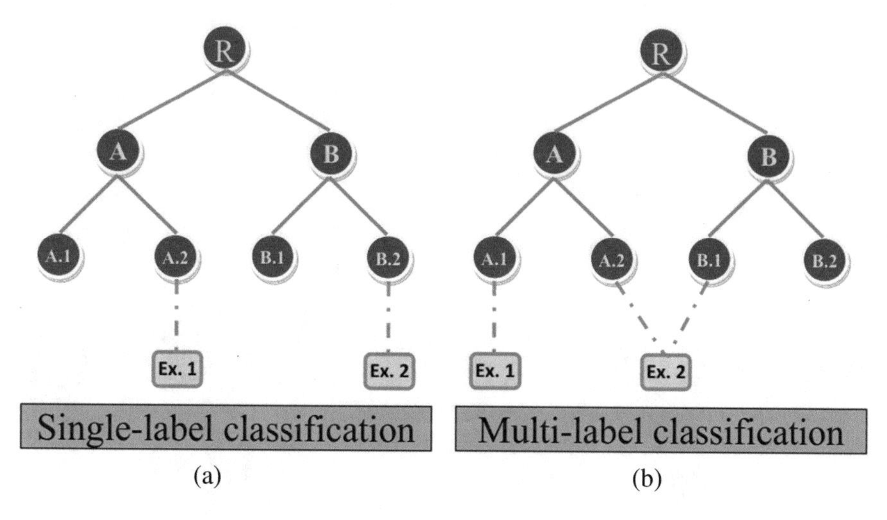

Fig. 1.3 Difference between single-label and multi-label classification. (**a**) Single-label classification: All instances are assigned to single class only (**b**) Multi-label classification: Instances may belong to multiple classes (Ex. 2 in the figure)

Challenges

1. Single-label versus multi-label classification—In single-label classification, each instance exclusively belongs to only one (single) class, whereas in multi-label classification instance may belong to several classes (Fig. 1.3). While single-label classification is easier, multi-label classification is difficult because instance can be associated with multiple classes that belongs to completely different branches in the hierarchy.

> It should be noted that multi-class and multi-label classification are two different terminology. Multi-class classification refers to a classification task with more than two classes. For example, classify a set of flowers which may be lily, rose, or lotus. Multi-class classification makes the assumption that each instance is assigned to one and only one label: a flower can be either rose or a lily but not both at the same time. On the other hand, in multi-label classification each instance can be assigned to multiple categories. For example, a movie can be categorized as both comedy and romantic at the same time.

2. Mandatory leaf node versus internal node prediction—In HC, each instance can be associated with labels that is on a path from the root node to a leaf node (full-path prediction) or stop at an internal node (partial-path prediction [6]). In mandatory leaf node prediction, only full-path predictions are allowed

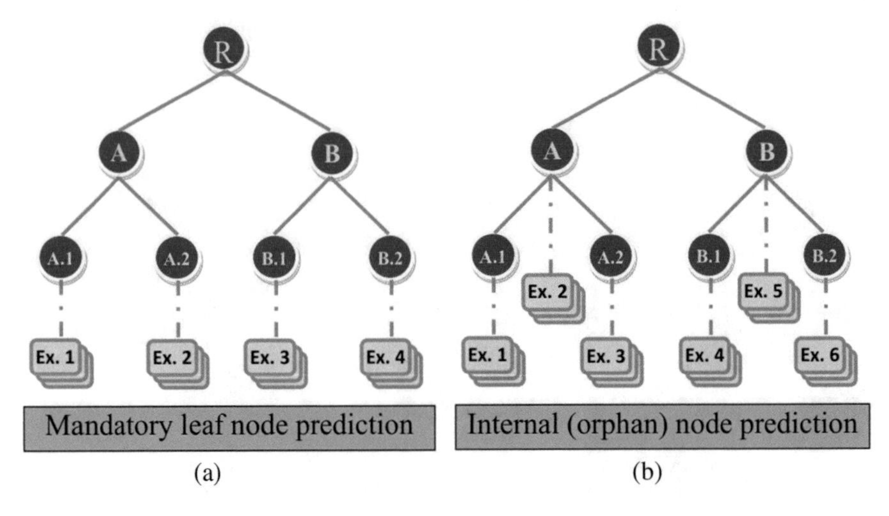

Fig. 1.4 Difference between mandatory leaf node and internal node prediction. (**a**) Mandatory leaf node prediction—All instances belong to leaf nodes. (**b**) Internal node prediction—Instances may belong to internal nodes (Ex. 2 and Ex. 5 in the figure)

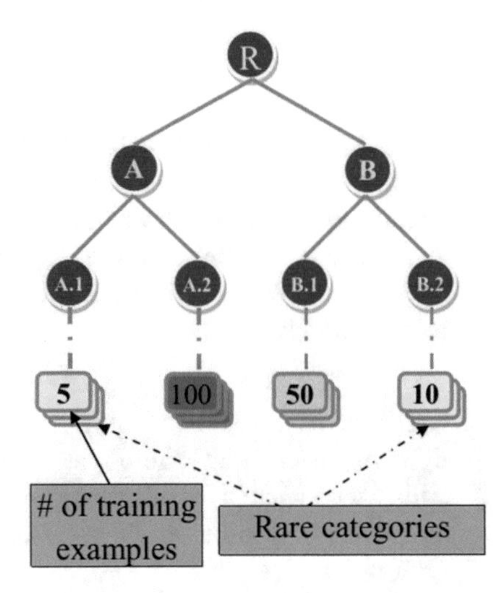

Fig. 1.5 Hierarchical structure showing classes with rare categories

for all instances (or in other words each instance is essentially assigned to leaf nodes in the hierarchy), whereas in internal node prediction, partial-path predictions are also allowed for instances (Fig. 1.4). Determining criterion for assigning instances to internal nodes is a difficult task. Internal node prediction is also referred to as nonmandatory leaf node prediction or orphan node detection problem in the literature [10].

3. Rare categories—In real-world problems, many classes having few positive instances which makes it difficult to learn generalized classification models as it is prone to over-fitting (Fig. 1.5). LSHC problem involves thousands of categories

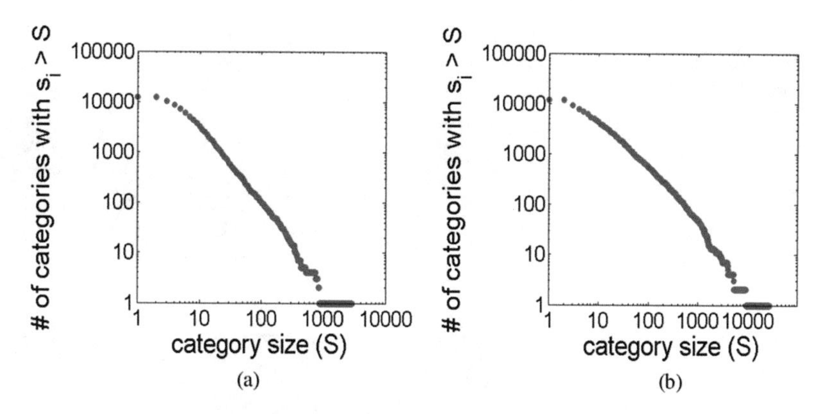

Fig. 1.6 Power law distribution followed in LSHTC datasets—(**a**) DMOZ-2010 and (**b**) DMOZ-2012

with varying distribution of instances per category. In datasets of such a large scale, skewed class distribution is observed where plenty of classes have fewer instances for training (e.g., more than 75% of LSHTC datasets belongs to rare categories, i.e., classes with less than ten examples) making it considerably difficult to learn a generalized model. This is known as the rare categories problem [1] and is more prominent in large-scale datasets because it exhibits power-law distribution for examples per class as shown in Eq. (1.1). To improve HC performance, it is necessary to address the issue of rare categories.

$$P(s_i > S) \propto S^{-\gamma} \tag{1.1}$$

where s_i denotes the size of i-th class and γ denotes the power law exponent. Figure 1.6a and b shows the power-law distribution followed in large-scale DMOZ-2010 and DMOZ-2012 datasets, respectively.

Although data size is increasing continuously, rare categories (or data sparsity) issue remains because more classes are also being added to the hierarchy.

Hierarchical structures are useful for improving performance on rare categories because they can exploit instances/parameters from parents/siblings relationships.

4. Feature selection—Real-world datasets often contain features that are either redundant (providing no useful information) or irrelevant (not able to discriminate between classes) for classification. Identifying discriminative features is crucial because not only does it improves classification performance but also improves runtime performance and memory required to store learned models.

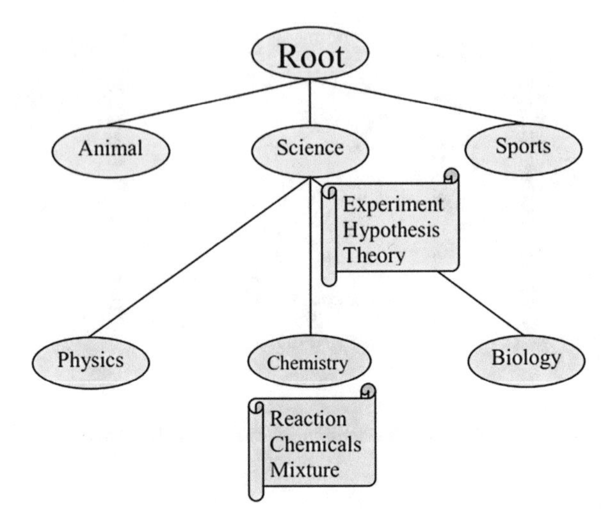

Fig. 1.7 Figure showing useful discriminative features for node—Science and Chemistry

However, feature selection is not an easy task, given large-scale datasets are high-dimensional and have hundreds and thousands of features. Figure 1.7 shows the importance of feature selection for HC. For discriminating *Science* from *Animal* and *Sports* class, features like *Experiment, Hypothesis*, and *Theory* are important, whereas features like *Computer, Flowers*, and *Vehicle* are irrelevant. Similarly, for discriminating *Chemistry* from *Physics* and *Biology* class, *Reaction, Chemicals, Mixture* are important features.

Features that are useful for discriminating parent node are not necessarily useful for discriminating children nodes. For example, in Fig. 1.7 although *Experiment, Hypothesis, Theory* are discriminating features for parent node *Science*, they are not useful for discriminating children node *Physics, Chemistry, Biology* because they are common features across all children.

5. Learning with hierarchical relationships—In order to learn generalized model, we need to incorporate hierarchical relationships information while training models (or classifiers). Parent-child and sibling relationships are widely used in literature to improve HC performance. However, incorporating relationships into model learning is nontrivial, and it may lead to optimization issues where finding the best solution might be a difficult task.
6. Scalability—Large-scale datasets are characterized by huge number of classes, features, and instances. Table 1.1 shows different data characteristics about DMOZ and ImageNet datasets. In order to deal with large-scale datasets, we need algorithms that are easily scalable by executing in distributed fashion or utilizing parallelized framework. Due to hierarchical dependencies between different classes, it is difficult to embarrassingly parallelize model learning,

Table 1.1 Large-scale dataset characteristics

Dataset	#Training examples	#Leaf node (classes)	#Features	#Parameters	Parameter size (approx)
DMOZ-2010	128,710	12,294	381,580	4,652,986,520	18.5 GB
DMOZ-2012	383,408	11,947	348,548	4,164,102,956	16.5 GB
ImageNet-2012	1.3M	1000	150,528	150,528,000	602 MB

causing runtime performance issue, and it may take several weeks or even months to train the models.

7. Hierarchical structure inconsistency—Hierarchies are designed by experts based on the domain knowledge. But in many cases, hierarchies are not suitable for classification due to presence of inconsistent parent-child/siblings relationships in the hierarchy. To improve performance we need to restructure the hierarchy to make it more consistent for classification.

   Designing a consistent hierarchy is challenging either due to insufficient domain knowledge or several confounding classes (such as *soc.religion.christian* and *talk.religion.misc* classes in Newsgroup dataset,[6] both relate to religion). Moreover, hierarchy generation based on semantics is susceptible to inconsistencies [8, 9]. This problem is more common for large-scale datasets. To illustrate in detail, consider the example shown in Fig. 1.8a, it consist of 1000 points divided into five classes that are generated using Gaussian distribution with different mean and variance. Figure 1.8b and c shows two different possible hierarchical structures with these classes. *Hierarchy 1* separates examples into two categories at level 1, namely, $\{(\blacklozenge, \blacksquare, \bigstar),(\blacktriangle, \bullet)\}$ which are not consistent for classification since category ($\bigstar$) is not easily separable from ($\blacktriangle$, $\bullet$) (assuming linear separators), whereas *Hierarchy 2* is more consistent since it groups easily separable classes together. This explains intuitively that inconsistent hierarchical structure can deteriorate the HC performance. To overcome this problem, it is necessary to develop methods that remove inconsistencies in the hierarchy prior to learning models.

> Resolving inconsistencies in the predefined (or original or expert defined) hierarchy serves as the preprocessing step, and any state-of-the-art existing HC algorithms can be applied on the modified hierarchy.

8. Error propagation—HC rely on several decisions that are made during the prediction phase. A number of decisions made are equivalent to path from root to the predicted leaf category. If incorrect decision is made at any point in the predicted path, it results in error being propagated at the lower levels

[6] http://qwone.com/~jason/20Newsgroups.

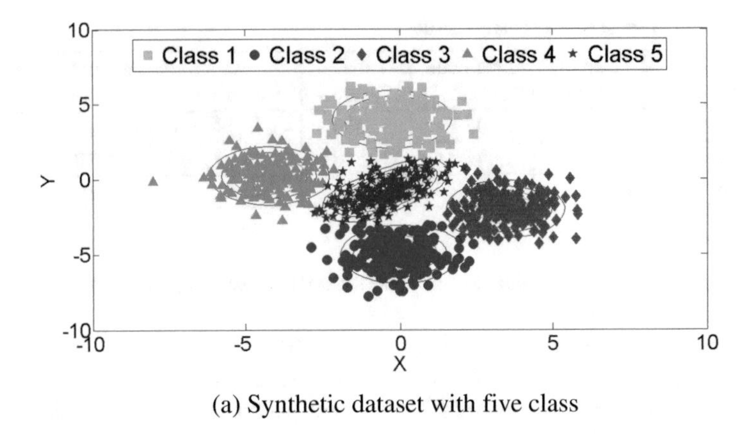

(a) Synthetic dataset with five class

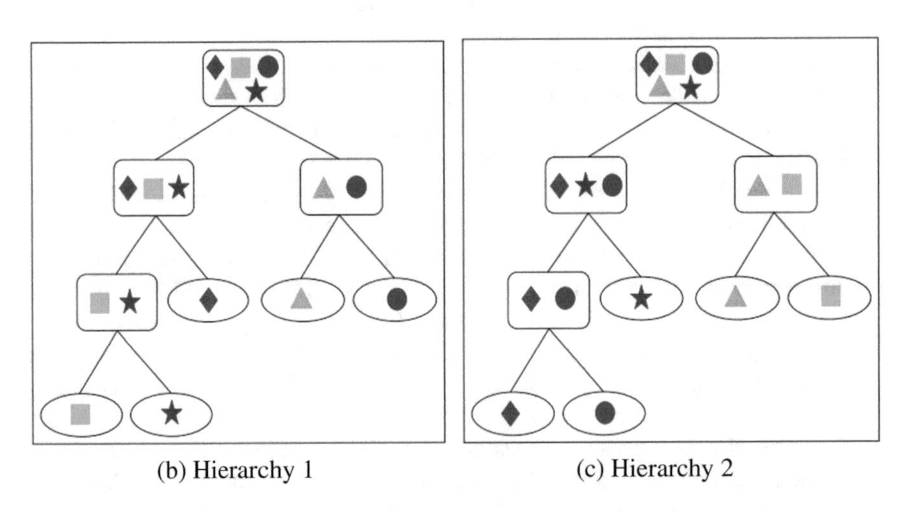

(b) Hierarchy 1 (c) Hierarchy 2

Fig. 1.8 (**a**) Synthetic dataset with five classes (marked with different symbol and color) and two different hierarchical structure shown in (**b**) and (**c**). *Hierarchy* 2 is better suited for classification as compared to *Hierarchy* 1 because in *Hierarchy* 2 classes are easily separable (assuming linear classifiers) at each level in the hierarchy

finally resulting in erroneous prediction. This phenomenon is known as error propagation and is most common in LSHC. Flat classifiers, on the other hand, rely on a single decision and are not prone to error propagation (even though the decision to be made is harder).

Due to the above challenges, the LSHC problem, in its most general form, is not easy to solve. In fact, most of the existing LSHC techniques solve a specific formulation of the problem. The formulation is induced by various factors such as number of rare categories, dimension of instances, etc.

1.3 Summary of Chapters

Chapter 2 discuss about various symbols and notations used in this book and provides an overview of the various existing HC methods in the literature. In Chap. 3, various hierarchical structure modification approaches to resolve inconsistencies within the hierarchy definition have been discussed followed by Chap. 4 that provides insight into different approaches for integrating information theoretic feature selection methods into the HC framework. In Chap. 5, we discuss various multi-task learning approaches for integrating information from multiple hierarchies. Finally, we conclude and provide various future research directions in Chap. 6.

References

1. Babbar, R., Partalas, I., Gaussier, E., Amini, M.R.: On flat versus hierarchical classification in large-scale taxonomies. In: Advances in Neural Information Processing Systems, pp. 1824–1832 (2013)
2. Bennett, P.N., Nguyen, N.: Refined experts: improving classification in large taxonomies. In: Proceedings of the 32nd international ACM SIGIR conference on Research and development in information retrieval, pp. 11–18 (2009)
3. Burred, J.J., Lerch, A.: A hierarchical approach to automatic musical genre classification. Citeseer
4. Cai, L., Hofmann, T.: Hierarchical document categorization with support vector machines. In: Proceedings of the thirteenth ACM International Conference on Information and Knowledge Management, pp. 78–87 (2004)
5. Cesa-Bianchi, N., Gentile, C., Zaniboni, L.: Hierarchical classification: combining bayes with svm. In: Proceedings of the 23rd International Conference on Machine Learning (ICML), pp. 177–184 (2006)
6. Cesa-Bianchi, N., Gentile, C., Zaniboni, L.: Incremental algorithms for hierarchical classification. Journal of Machine Learning Research **7**(Jan), 31–54 (2006)
7. Gopal, S., Yang, Y.: Recursive regularization for large-scale classification with hierarchical and graphical dependencies. In: Proceedings of the 19th ACM SIGKDD International Conference on Knowledge Discovery and Data mining, pp. 257–265 (2013)
8. Naik, A., Rangwala, H.: Filter based taxonomy modification for improving hierarchical classification. http://arxiv.org/abs/1603.00772 (2016)
9. Naik, A., Rangwala, H.: Inconsistent node flattening for improving top-down hierarchical classification. In: IEEE International Conference on Data Science and Advanced Analytics (DSAA), pp. 379–388 (2016)
10. Naik, A., Rangwala, H.: Integrated framework for improving large-scale hierarchical classification. In: 16th IEEE International Conference on Machine Learning and Applications (ICMLA), pp. 281–288 (2017)
11. Silla Jr, C.N., Freitas, A.A.: A survey of hierarchical classification across different application domains. Data Mining and Knowledge Discovery **22**(1–2), 31–72 (2011)
12. Xue, G.R., Xing, D., Yang, Q., Yu, Y.: Deep classification in large-scale text hierarchies. In: Proceedings of the 31st annual International ACM SIGIR Conference on Research and Development in Information Retrieval, pp. 619–626 (2008)
13. Zhou, D., Xiao, L., Wu, M.: Hierarchical classification via orthogonal transfer. In: Proceedings of the 28th International Conference on Machine Learning (ICML), pp. 801–808 (2011)

Chapter 2
Background

This chapter provides a description of the various symbol notations used in this book. We have used consistent symbolic notations across all chapters for easily grasping the concepts. We also discuss about different methods for solving HC problem. Many advanced work has been accomplished by the researchers in the field of HC. However, we will restrict our discussion to those methods that are useful and widely popular including some of the methods that achieves state-of-the-art results on large-scale datasets. Finally, we also touch base upon various available hierarchical datasets and different evaluation metrics used for comparing the hierarchical classifiers.

2.1 Notations

This section discusses the commonly used notations in this book. Summary of notations are described in Table 2.1. For ease of understanding, vectors and matrices are denoted by **bold** letters. Moreover, for representing matrices we have used uppercase letters and vectors are represented using lowercase letters.

Specific to the HC problem, $\mathcal{N}$ denotes the total number of nodes in the hierarchy. Total number of training instances is denoted by symbol N, where N input training pairs are represented using $\mathbf{D} = \{(\mathbf{x}(i), y(i))\}_{i=1}^{N}$. $\mathbf{x}(i) \in X$ corresponds to the i-th input vector in the input domain (space) X and $y(i) \in Y$ corresponds to the true label in the output domain (space) Y. L denotes the total number of leaf nodes in the hierarchy.

For binary classifiers, learned optimal model weight vectors corresponding to the n-th node in the hierarchy H is represented using $\mathbf{w}_n$. Group of m weight vectors are represented using the notation $[\mathbf{W}]_{m*d}$, where d corresponds to the dimensionality (number of features) of the input vector. For multi-class classifiers, multiple classifiers are trained at each of the internal node in the hierarchy.

A. Naik, H. Rangwala, *Large Scale Hierarchical Classification: State of the Art*, SpringerBriefs in Computer Science, https://doi.org/10.1007/978-3-030-01620-3_2

Table 2.1 Notations

Symbols	Description
$\mathbb{R}$	Set of real numbers
H	Original given hierarchy
N	Total number of training examples
$\mathcal{N}$	Total number of nodes in the hierarchy
d	Dimensionality (number of feature) of input vector
L	Set of leaf categories (classes or labels)
$\mathbf{X}$	Input domain (space)
$\mathbf{Y}$	Output/label domain (space)
$\mathbf{x}(i) \in \mathbb{R}^d$	Input vector for i-th training example where k-th feature is denoted by $\mathbf{x}_k(i)$
$y(i) \in L$	True label for i-th training example
$\mathbf{w}_n$	Learned model weight vectors using binary classifiers for n-th node in the hierarchy
$\mathbf{w}_n^c$	Learned model weight vectors using multi-class classifiers corresponding to c-th children of n-th node in the hierarchy (unless stated explicitly)
$C(n)$	Set of children for n-th node in the hierarchy
$\pi(n)$	Parent of n-th node in the hierarchy
$S(n)$	Set of siblings for n-th node in the hierarchy
$A(n)$	Set of n-th node ancestors including the node itself but excluding root
$\widehat{\mathbf{x}}(i) \in \mathbb{R}^d$	Input vector for i-th test example
$\widehat{y}(i) \in L$	Predicted label for i-th test example
$y_n(i) \in \pm 1$	Binary label used for i-th training example to learn weight vectors for n-th node in the hierarchy, $y_n(i) = 1$ iff $y(i) = n$, -1 otherwise
$y_n^c(i) \in \pm 1$	Binary label used for i-th training example to learn weight vector corresponding to c-th child of n-th node in the hierarchy, $y_n^c(i) = 1$ iff $y(i) = c$, -1 otherwise
Ψ_n	Optimal objective function value for n-th node in the hierarchy
Ψ_n^c	Optimal objective function value for c-th child of n-th node in the hierarchy

Therefore, we use the notation $\mathbf{w}_n^c$ to represent the learned model corresponding to c-th child of node n, whereas combined model at node n is represented using $\mathbf{W}_n = [\mathbf{w}_n^c]_{c \in C(n)}$. $C(n)$ denotes the set of n-th node children. Parent and siblings of node n are denoted by symbols $\pi(n)$ and $S(n)$, respectively. $A(n)$ denotes the set of n-th node ancestors including the node itself but excluding root.

For training binary classifiers at node n, we use the binary label $y_n(i) = \pm 1$ for i-th training instance where $y_n(i) = 1$ iff $y(i) = n$ and -1, otherwise. Similarly, for multi-class classifiers, we use the binary label $y_n^c(i) = \pm 1$ for i-th training instance where $y_n^c(i) = 1$ iff $y(i) = c$ and -1, otherwise. Predicted label for the i-th test instance $\widehat{\mathbf{x}}(i)$ is represented using the notation $\widehat{y}(i)$. Further, Ψ_n and Ψ_n^c denote the optimal objective function value for n-th node and its c-th children in the hierarchy, respectively.

2.2 Different Approaches for Hierarchical Classification

HC is one of the most important problems in data mining and machine learning community that has received significant interest among researchers due to its practical importance (evident from various online large-scale hierarchical classification challenges such as LSHTC, BioASQ, ImageNet). In general, HC problem can be formally defined as:

Definition 2.1 (Problem Definition) Given a hierarchy H defined over the output (label) space Y and a set of N training examples composed of pairs $D=\{(\mathbf{x}(i), y(i))\}_{i=1}^{N}$, where $(\mathbf{x}(i), y(i)) \in X \times Y$, the goal of the hierarchical classification is to learn a mapping function $f : X \in \mathbb{R}^d \rightarrow Y$ that maps the inputs in the input space X to outputs in the output space Y, such that the function f is accurately able to predict the output y of an input instance $\mathbf{x}$ and generalizes well to data that is not observed during the training.

HC is the process of classifying unlabeled instances into a hierarchical organization of classes.

In literature various methods exist to solve the HC problem based on how the hierarchical relationships information is leveraged during the model learning [34]. One of the simplest approaches, known as flat classification, disregards the hierarchical structure and train classifiers for each of the leaf node (categories) to discriminate from remaining leaf nodes. Other approaches involve utilizing the hierarchical relationships information during the learning and/or predicting phase, e.g., local and global classifiers. While local classifiers are trained by splitting the hierarchical structure into several smaller structures for utilizing the local relationships, global classifiers are trained by considering the entire class hierarchy at once. In general, HC approaches can be broadly divided into three categories—*flat*, *local*, and *global* classification approaches. Figure 2.1 provides an overview of different HC approaches. More details about these approaches are discussed in the next few subsections.

2.2.1 *Flat Classification Approach*

This is one of the simplest and straightforward implementations of the standard classification algorithm into the HC problem. In this method, we ignore the hierarchy and train an independent one-vs-rest binary or multi-class classifiers corresponding to each of the leaf categories that can discriminate it from remaining

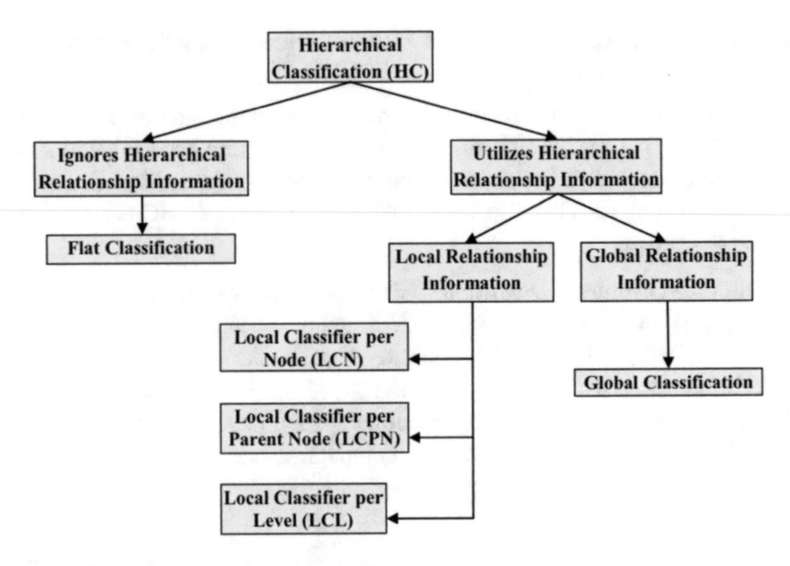

Fig. 2.1 Different approaches for solving HC problem

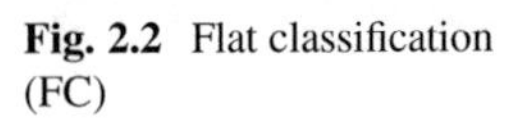

Fig. 2.2 Flat classification (FC)

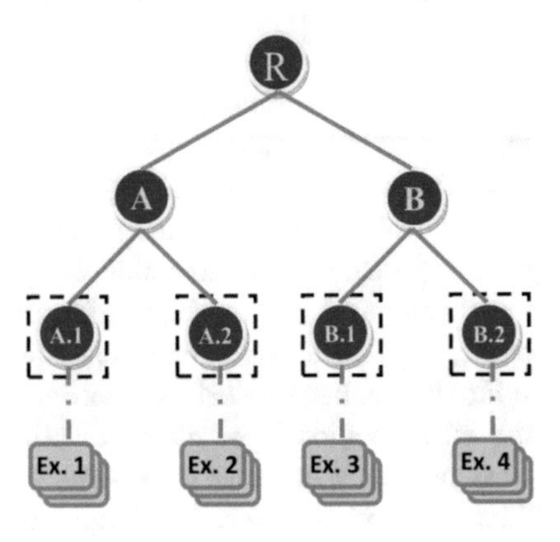

leaf categories as shown in Fig. 2.2. Label prediction $\widehat{y}$ for an unknown test instance $\mathbf{x}$ is done according to the rule shown in Eq. (2.1):

$$\widehat{y} = \underset{y \in Y}{\mathbf{argmax}} \ f(\mathbf{x}, y|\mathbf{w}) \tag{2.1}$$

where the function $f : X \rightarrow Y$ is parameterized by the model weight vector $\mathbf{w}$.

This approach provides an indirect solution to the HC problem because all the ancestors associated with the predicted leaf category are also assigned to the test instance. For example, in Fig. 2.2 an instance classified as $B.1$ also belongs to node B and R. Some of the well-known standard formulation of binary [21] and multi-class [10] classifiers that can be used for flat classification are shown in Eqs. (2.2) and (2.3):

$$\textbf{Binary classifier} \quad \underset{\mathbf{w}_l}{\textbf{minimize}} \sum_{i=1}^{N} \mathscr{L}\big(\mathbf{w}_l, \mathbf{x}(i), y(i)\big) + \lambda \big|\big|\mathbf{w}_l\big|\big|_2^2 \tag{2.2}$$

$$\textbf{Multi-class classifier} \quad \underset{\{\mathbf{w}_l\}_{l \in L}, \{\xi_i\}_{i=1}^{N}}{\textbf{minimize}} \frac{1}{N} \sum_{i=1}^{N} \xi_i + \lambda \sum_{l=1}^{L} \big|\big|\mathbf{w}_l\big|\big|_2^2 \tag{2.3}$$

$$\textbf{s.t.}: \mathbf{w}_{l_i}^T \mathbf{x}(i) - \mathbf{w}_l^T \mathbf{x}(i) \geq 1 - \xi_i,$$

$$\forall l \in L - \{l_i\}, \forall i \in [1, \cdots, N]$$

$$\xi_i \geq 0, \forall i \in [1, \cdots, N]$$

where $\lambda > 0$ is the penalty parameter, $\mathscr{L}$ denotes the loss function such as hinge loss or logistic loss, ξ_i denotes the slack variables, and $|| \cdot ||_2^2$ denotes the squared l_2-$norm$.

Although flat classification approach is known for its simplicity and has been shown to work well in practice for small and well-balanced datasets [43], its performance suffers when the number of classes (categories) that needs to be discriminated becomes huge and is not balanced [2], potentially containing lots of rare categories. It also has a major problem with longer training and prediction time because it considers all the examples during the model training and invokes all the models for the leaf categories to make label prediction making it computationally expensive, especially for large-scale datasets. Flat approach also assumes that all instances belong to leaf categories, whereas in many real-world applications, instances don't necessarily belong to leaf categories (nonmandatory leaf node prediction problem). For example, given a hierarchy containing a path $Root \rightarrow Science \rightarrow Medicine$, an article discussing about $Science$ in general should only be categorized as $Root \rightarrow Science$ and not $Medicine$. Another problem with flat approach is with multi-label classification where instances can belong to several different categories. In such case, flat classification becomes inefficient and hierarchical classification is preferred.

2.2.2 *Local Classification Approach*

This method explores local hierarchical structure information such as parent-child and siblings relationships during the model learning. Based on how the local information are extracted during the model learning, local classification approach can be further categorized into three broad categories—*local classifier per node* (LCN), *local classifier per parent node* (LCPN), and *local classifier per level* (LCL).

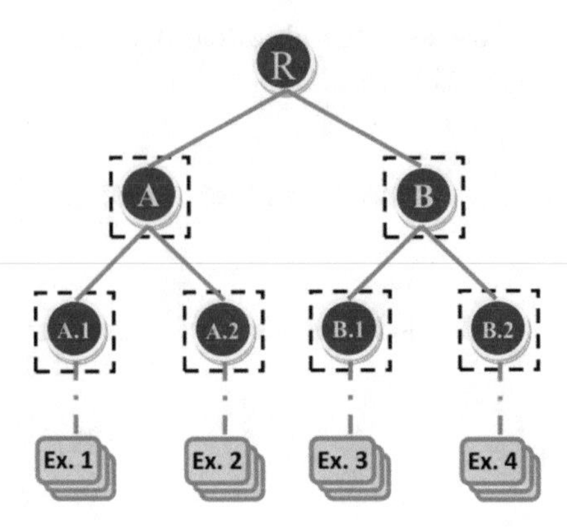

Fig. 2.3 Local classifier per node (LCN)

2.2.2.1 Local Classifier per Node Approach

In this approach, binary classifier Ψ_n is learned for each node $n \in \mathcal{N}$ (except root) in the hierarchy H as shown in Fig. 2.3. The dashed squares in the figure represent binary classifiers. Goal of this approach is to learn the classifiers that can effectively discriminate between the sibling nodes in the hierarchy. Usually, for training the classifier at a node, we assign examples belonging to the n-th node and its descendants as the positive training examples and those belonging to the siblings of the n-th node and their descendants as the negative examples. However, in literature different criteria for defining the positive and negative examples have been used [5, 15, 16].

To make the label prediction of an unknown test instance $\mathbf{x}$, the algorithm (shown in Eq. (2.9)) typically proceeds in the top-down fashion starting at the root and recursively selecting the best children till it reaches a terminal node that belongs to the set of leaf categories L, which is the final predicted label.

This approach although popular in literature suffers from training a large number of binary classifiers. Other problem includes inconsistent predictions obtained by the set of binary classifiers which lead to horizontal and vertical inconsistencies [32]. Therefore, LCN approach must be coupled with a method to avoid horizontal and vertical inconsistency.

Horizontal Inconsistency—Inconsistent predictions made by classifiers where instance may be associated with different nodes in the same level. For example, in Fig. 2.3 horizontal inconsistent prediction occurs if the prediction made by classifiers is A, $A.1$, and $A.2$.

Vertical Inconsistency—Inconsistent predictions made by classifiers where instance may be associated with different branches in the hierarchy. For example, in Fig. 2.3 vertical inconsistent prediction occurs if the prediction made by classifiers is A, $A.1$, and $B.1$.

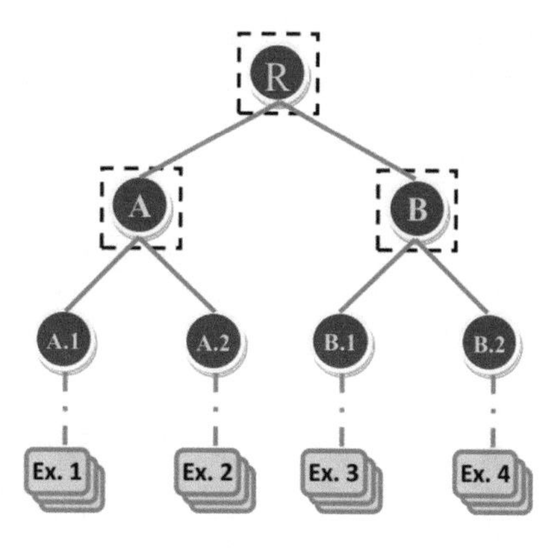

Fig. 2.4 Local classifier per parent node (LCPN)

2.2.2.2 Local Classifier per Parent Node Approach

In this approach, multi-class classifier is learned for each of the parent nodes in the hierarchy H as shown in Fig. 2.4. The dashed square in the figure represents multi-class classifiers. Like local classifier per node approach, the goal of this approach is to learn classifiers that can effectively discriminate between the siblings. For training the classifier at each parent node p, we use the examples from its descendants where each of the children categories $C(p)$ of parent node p corresponds to different classes. Predicting the label for an unknown test instance $\mathbf{x}$ is done in a similar manner as shown in Eq. (2.9).

2.2.2.3 Local Classifier per Level Approach

In this approach, multi-class classifier is learned for each level in the hierarchy as shown in Fig. 2.5. Among local approaches, this is the least popular approach in the literature. For training the classifier at each level, we use the examples from nodes in the level and its descendants, where different nodes in the level correspond to different classes. Prediction for an unknown test instance $\mathbf{x}$ is done by choosing the best node at each level in the hierarchy H. Since classifiers at each level makes independent predictions, there is possibility that this approach may lead to vertical inconsistency prediction. For example, in Fig. 2.5 inconsistent prediction occurs if the prediction made by the level 1 classifier is B, whereas the level 2 classifier predicts A.2 that corresponds to different branch in the hierarchy. In order to make this approach useful, classification results are complemented with a post-processing step to resolve such inconsistent predictions. Paes et al. [32] discusses different approaches for dealing with inconsistencies in LPL.

2.2.3 *Global Classification Approach*

This approach is often referred as the *big-bang* approach in the literature [9]. Unlike local approaches, global classification approach learns a single complex

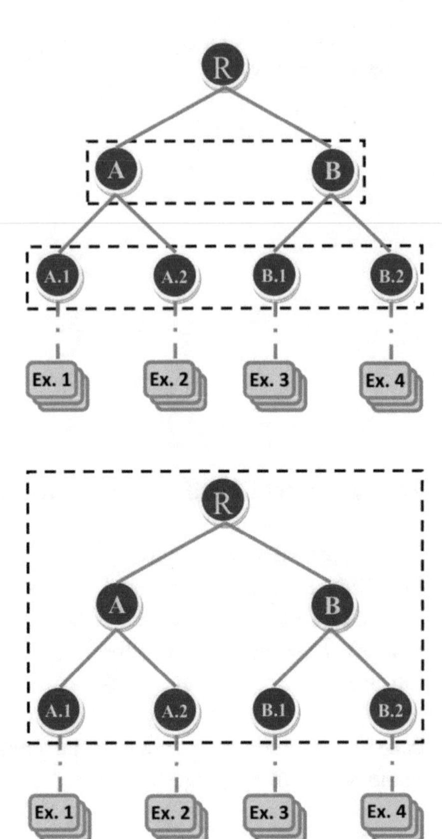

Fig. 2.5 Local classifier per level (LPL)

Fig. 2.6 Global classification (GC)

classification model by taking into account the class hierarchy as a whole as shown in Fig. 2.6. For predicting the labels of an unknown test instance $\mathbf{x}$, an approach similar to flat or local methods is followed.

The main advantage of this approach is that there is no need to train a large number of classifiers. This approach also doesn't require to deal with the inconsistency in the prediction of classes. Its main disadvantage is the increased complexity on learning the global classifier.

2.3 Model Learning: General Formulation

In classification, models are learned by combining two terms—empirical loss and regularization:

1. Empirical Loss—It controls how well the learned models fit the training data.
2. Regularization—It prevents models from over-fitting and encodes additional information such as hierarchical relationships.

Generally, model learning can be represented using Eq. (2.4):

$$\underset{\mathbf{W}}{\textbf{minimize}} \mathscr{L}\big(f(\mathbf{X}, \mathbf{W}), \mathbf{Y}\big) + \lambda \Omega\big(\mathbf{W}\big) \tag{2.4}$$

where $\mathscr{L}(.)$ is the loss function and $\Omega(.)$ denotes the regularization. $\lambda > 0$ is the hyper-parameter that controls trade-off between loss function and regularization term. In literature, hinge loss and logistic loss are widely used for classification. Equations (2.5) and (2.6) show the hinge loss and logistic loss, respectively. For regularization, $l_1\text{-}norm$ and $l_2\text{-}norm$ are popular. It is computed as shown in Eqs. (2.7) and (2.8). While $l_1\text{-}norm$ enforces sparsity by only selecting relevant features, $l_2\text{-}norm$ squeezes the model parameters to be close to 0, thereby preventing parameters from taking extremely larger values:

$$Hinge\ loss = \textbf{max}\big(0, 1 - yf(\mathbf{x}, \mathbf{w})\big) = \big|1 - yf(\mathbf{x}, \mathbf{w})\big|_{+} \tag{2.5}$$

$$Logistic\ loss = \frac{1}{ln\ 2} ln\Big(1 + exp\big(-yf(\mathbf{x}, \mathbf{w})\big)\Big) \tag{2.6}$$

$$l_1\text{-}norm = \big|\big|\mathbf{W}\big|\big|_1 = |\mathbf{w}_1| + |\mathbf{w}_2| + \cdots + |\mathbf{w}_d| \tag{2.7}$$

$$l_2\text{-}norm = \big|\big|\mathbf{W}\big|\big|_2 = \sqrt{\mathbf{w}_1^2 + \mathbf{w}_2^2 + \cdots + \mathbf{w}_d^2} \tag{2.8}$$

There are multiple ways of selecting positive and negative instances for training classifier at each node n in the hierarchy. We will discuss two of the most widely used method:

1. *One-versus-rest*—In this approach, all instances that belong to children nodes of node n are treated as a positive, whereas all other instances are considered as negative for training classifier at node n.
2. *One-versus-siblings*—In this approach, all instances that belong to children nodes of node n are treated as a positive, whereas instances from sibling classes *only* are considered as negative for training classifier at node n.

Figure 2.7 visually illustrates both these approaches where classifier is being trained at node *A.1*.

2.3.1 Top-Down Hierarchical Classification

One of the most efficient approach for solving LSHC problem is using the top-down methods [22, 26]. In this method, local or global classification method is used for model training, and the unlabeled instances are recursively classified in a top-down fashion as shown in Eq. (2.9). At each step, the best node is picked based on the computed prediction score of its children nodes. The process repeats until the leaf

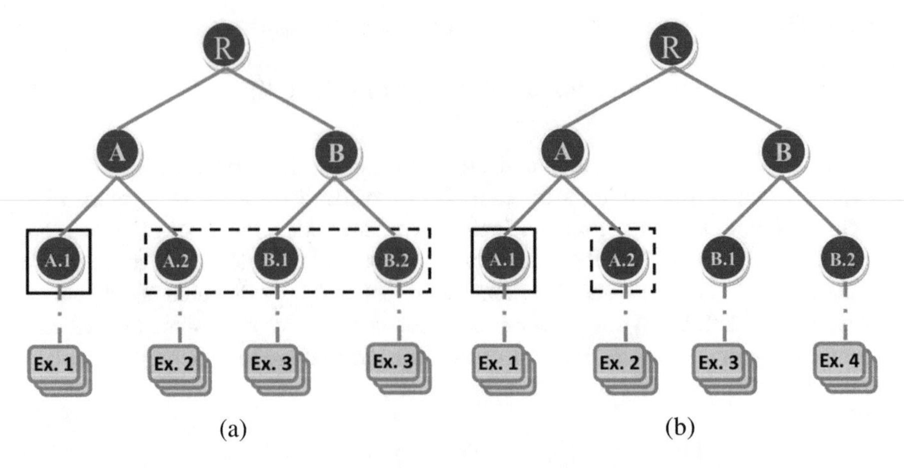

Fig. 2.7 Different ways of selecting positive and negative instances for training classifier at node A.1. Positive instances are shown by solid rectangle, whereas negative instances are shown by dotted rectangle. (**a**) One-versus-rest. (**b**) One-versus-siblings

node representing a certain category (or class label) is reached, which is the final predicted label.

For internal node prediction problem, predictive threshold can be defined at each node. If the probability of instance is greater than a defined predictive threshold (e.g., 0.8) at the node, then the instance is recursively classified to its lower-level child node for further prediction until threshold is violated or leaf node is reached.

Defining predictive threshold is a crucial task. Smaller threshold will result in instances wrongly classified into the lower levels in the hierarchy while larger threshold will block the instances at higher levels. SCut is one of the popular method for defining threshold [40].

$$\widehat{y} = \left\{ \begin{array}{l} \textbf{initialize} \quad n := Root \\ \textbf{while}\ n \notin L \\ \quad n := \textbf{arg max}_{q \in C(n)}\ f_q(\mathbf{x}) \\ \textbf{return}\ n \end{array} \right\} \tag{2.9}$$

Top-down (TD) methods are popular for large-scale problems owing to their computational benefits where only the subset of classes in the relevant path are considered during prediction phase. It also doesn't suffer from inconsistencies problem (horizontal, vertical) because only the best child node is selected at each path down the level. In the past, top-down methods have been successfully used to solve HC problems [14, 20, 25]. Liu et al. [26] performed classification on large-scale Yahoo! dataset and analyzed the complexity of the top-down approach. In Secker et al. [33], a selective classifier top-down method is proposed where the classifier to train at particular node is chosen in a data-driven manner.

It is important to note that for multi-label predictions, multiple paths can be followed in TD methods based on prediction probabilities.

2.4 Hierarchical Datasets

There are lot of hierarchical datasets available for testing the performance of HC. Range of datasets vary from shallow to deep hierarchies, few to high-dimensional features, balanced to skewed distribution, and text to image. In this section we discuss some of the datasets that are often used for evaluating the HC performance.

Datasets

1. NEWSGROUP (NG)[1]—It is a collection of approximately 20,000 news documents partitioned (nearly) evenly across 20 different topics such as *baseball*, *electronics*, and *graphics*.
2. CLEF [12]—Dataset contains medical images annotated with Information Retrieval in Medical Applications (IRMA) codes. Each image is represented by the 80 features that are extracted using local distribution of edges method. IRMA codes are hierarchically organized.
3. IPC[2]—Collection of patent documents organized in International Patent Classification (IPC) hierarchy. This is an example of imbalanced, high-dimensional dataset.
4. DIATOMS [13]—Diatom images that was created as the part of the ADIAC project. Features for each image is created using various feature extraction techniques mentioned in [13]. This dataset have examples that are assigned to internal nodes in the hierarchy.
5. RCV1 [24]—It is a multi-label text classification dataset extracted from Reuters Corpus of manually categorized newswire stories. This is an example of multi-label dataset.

[1] http://qwone.com/~jason/20newsgroups/.

[2] http://www.wipo.int/classifications/ipc/en/.

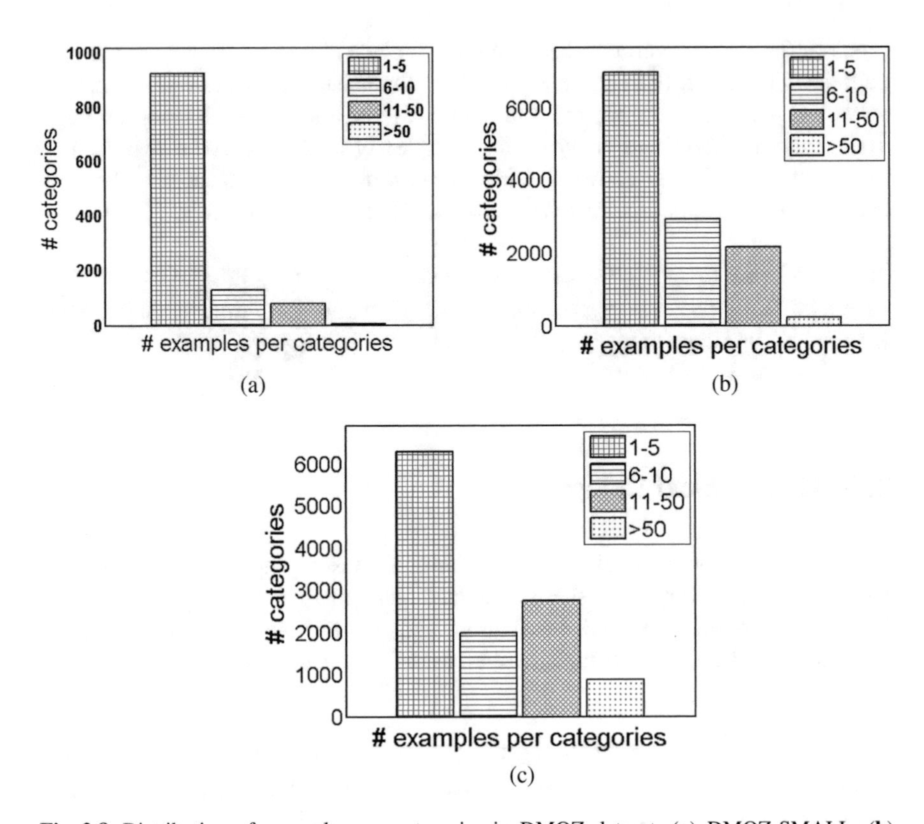

Fig. 2.8 Distribution of examples per categories in DMOZ dataset. (**a**) DMOZ-SMALL. (**b**) DMOZ-2010. (**c**) DMOZ-2012

6. LSHTC (DMOZ-SMALL, DMOZ-2010, and DMOZ-2012[3])—Multiple web documents organized in various classes using the hierarchical structure. Dataset has been released as the part of the PASCAL Large Scale Hierarchical Text Classification (LSHTC)[4] challenge in the year 2010 and 2012. DMOZ datasets are characterized by high-dimensional features with more than 75% of the classes belonging to rare categories (i.e., classes having $\leq$ 10 examples as shown in Fig. 2.8). Moreover, DMOZ-2010 and DMOZ-2012 are large-scale datasets with more than 10,000 classes. For all the datasets, the hierarchy has a tree structure and the internal nodes define a virtual category tree, i.e. the examples are assigned to only leaf nodes directly.

Table 2.2 shows various statistics about discussed datasets.

[3]http://www.dmoz.org/.

[4]http://lshtc.iit.demokritos.gr/.

Table 2.2 Dataset statistics

Dataset	Total nodes	#Categories	Levels	#Training instances	#Test instances	#Features	Avg. #labels per example
NG	28	20	3	11,269	7505	61,188	1
CLEF	88	63	3	10,000	1006	80	1
IPC	553	451	3	46,324	28,926	1,123,497	1
DIATOMS	399	311	3	3119	1054	371	1
RCV1	117	101	5	23,149	781,265	48,728	3.18
DMOZ-SMALL	2388	1139	5	6323	1858	51,033	1
DMOZ-2010	17,222	12,294	5	128,710	34,880	381,580	1
DMOZ-2012	13,963	11,947	5	383,408	103,435	348,548	1

Usually small percentage of training instances are used as the validation set. It is used to tune model parameters, select features, and make other decisions regarding the learning algorithm.

2.5 Evaluation Metrics for Hierarchical Classification

Accuracy is one of the widely used metrics for evaluating classifier performance. It is defined as the ratio of correct predictions and total predictions. The major disadvantage of using accuracy as the evaluation metric is that if the number of positive and negative examples is imbalanced, then accuracy doesn't provide reliable results. Classifier can achieve high accuracy by predicting all examples as the majority class. Real-world datasets are often imbalanced and therefore accuracy cannot be used. To overcome this problem, other measures that rely on precision and recall are often used. Below we will discuss some of the most widely used measures for evaluating the HC performance.

2.5.1 *Flat Measures*

Standard set-based measures [39] Micro-F_1 (μF_1) and Macro-F_1 (MF_1) are commonly used for evaluating the HC performance. To compute μF_1, we sum up the category-specific true positives (TP_c), false positives (FP_c), and false negatives (FN_c) for different categories and compute the μF_1 score as

$$P = \frac{\sum_{c \in L} TP_c}{\sum_{c \in L} (TP_c + FP_c)}$$

$$R = \frac{\sum_{c \in L} TP_c}{\sum_{c \in L}(TP_c + FN_c)}$$

$$\mu F_1 = \frac{2PR}{P + R} \tag{2.10}$$

Definition 2.2 True positive for category c (TP_c)—The number of examples that belongs to category c and correctly predicted by model.

Definition 2.3 False positive for category c (FP_c)—The number of examples that doesn't belong to category c and incorrectly predicted by model.

Definition 2.4 False negative for category c (FN_c)—The number of examples that belongs to category c and incorrectly predicted by model.

Unlike μF_1, $\mathrm{M}F_1$ gives equal weightage to all the categories so that the average score is not skewed in favor of the larger categories. $\mathrm{M}F_1$ is defined as follows:

$$P_c = \frac{TP_c}{TP_c + FP_c}$$

$$R_c = \frac{TP_c}{TP_c + FN_c}$$

$$MF_1 = \frac{1}{|L|} \sum_{c \in L} \frac{2P_c R_c}{P_c + R_c} \tag{2.11}$$

where, $|L|$ is the number of categories (classes).

Flat measures, although useful, suffer from one major drawback. As shown in Fig. 2.9, in flat measures all misclassified examples are treated equally while evaluating the performance metric. Intuitively, for HC this doesn't make much sense because degree of misclassification is an important factor to consider while evaluating HC performance. Models doing misclassification into neighborhood classes are comparatively better than ones misclassifying into farther classes.

2.5.2 *Hierarchical Measures*

Different from flat measures that penalize each of the misclassified examples equally, hierarchical measures take into consideration hierarchical distance between the true and predicted label for evaluating the classifier performance [23]. Misclassifications that are closer to the actual class are less severe than misclassifications that are farther from the true class with respect to the hierarchy (e.g., instance from *Hockey* class misclassified as the *Baseball* class is less severe in comparison to the *Hockey* misclassified as *Cat* because *Hockey* and *Baseball* class belongs to common parent category *Sports*, whereas *Cat* class which is child of *Animal*

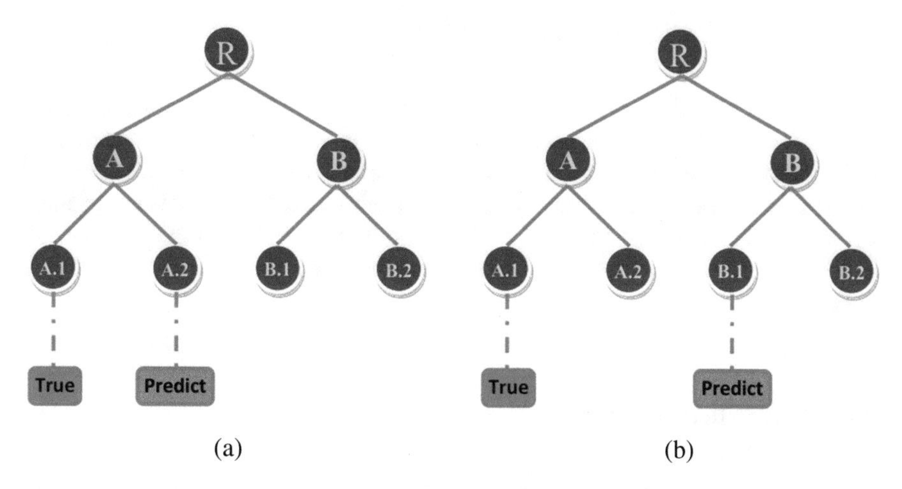

Fig. 2.9 Both misclassification are treated equally in flat measures. (**a**) Misclassification 1. (**b**) Misclassification 2

class belongs to completely different branch in the hierarchy). The hierarchy-based measures include hierarchical F_1 (hF_1) (harmonic mean of hierarchical precision (hP), hierarchical recall (hR)) and tree-induced error (TE) [11], which are defined as follows:

$$hP = \frac{\sum_{i=1}^{N} \left| A\left(\widehat{y}(i)\right) \cap A\left(y(i)\right) \right|}{\sum_{i=1}^{N} \left| A\left(\widehat{y}(i)\right) \right|}$$

$$hR = \frac{\sum_{i=1}^{N} \left| A\left(\widehat{y}(i)\right) \cap A\left(y(i)\right) \right|}{\sum_{i=1}^{N} \left| A\left(y(i)\right) \right|}$$

$$hF_1 = \frac{2 * hP * hR}{hP + hR} \tag{2.12}$$

$$TE = \frac{1}{N} \sum_{i=1}^{N} \delta\left(\widehat{y}(i), y(i)\right) \tag{2.13}$$

where $A\left(\widehat{y}(i)\right)$ and $A\left(y(i)\right)$ are, respectively, the sets of ancestors of the predicted and true label which include the label itself, but do not include the root node. $\delta(a, b)$ gives the length of undirected path between categories a and b in the hierarchy.

Tree-induced error is an error measure—the lower its value, the better the model.

2.5.3 *Area Under the Curve (AUC)*

It is used in classification analysis to determine which of the evaluated models predict the classes best. An example of AUC is receiver operating characteristic (ROC) curve and precision-recall curve. ROC curve is widely used and it is created by plotting the true positive rates against false positive rates. AUC is then computed by taking area under the ROC curve [35]. The closer AUC for a model comes to 1, the better it is. So models with higher AUC scores are preferred over those with lower AUC scores.

2.6 Literature Review

There have been numerous work proposed in the literature to address the problem of HC. Table 2.3 shows the summarized snapshot of different methods. In this section, we will review in detail some of the most popular methods.

2.6.1 *Hierarchical Orthogonal Transfer*

Classifying sibling classes at lower level becomes difficult as the depth of the hierarchy increases. This is because classes become more similar to each other. To address this problem, a hierarchical SVM learning formulation has been proposed by Zhou et al. in the paper [41]. It enforces the learned model parameters of each node to be orthogonal to all its ancestor nodes at higher levels. Orthogonality between the i-th node and its ancestor nodes $j \in A(i)$ is incorporated by introducing

Table 2.3 Summarization of different HC methods categorized by issue addressed

Reference	Issue addressed
Zhou et al. [41], Gopal et al. [19]	Learning with hierarchical relationships
McCallum et al. [27], Naik et al. [28]	Rare categories
Xue et al. [38], Anveshi et al. [7], Gopal et al. [17]	Scalability
Anveshi et al. [8]	Scalability, rare categories, multi-label
Naik et al. [29], Babbar et al. [1]	Scalability, feature selection
Gopal et al. [18]	Scalability, learning with hierarchical relationships multi-label
Naik et al. [30, 31], Rohit et al. [3]	Hierarchical inconsistency
Bennett et al. [4], Zhu et al. [42]	Error propagation
Cesa-Bianchi et al. [6], Vens et al. [37]	Multi-label

the regularization constraints $|\mathbf{w}_i^T \mathbf{w}_j|$ into the minimization objective function. Optimization formulation for their proposed approach can be represented using Eq. (2.14):

$$\begin{aligned} \textbf{minimize}: &\quad \frac{1}{2}\sum_{i,j=1}^{\mathcal{N}} \boldsymbol{K}_{ij}|\mathbf{w}_i^T \mathbf{w}_j| + \frac{\lambda}{N}\sum_{k=1}^{N} \xi_k \\ \textbf{subject to}: &\quad \mathbf{w}_i^T \mathbf{x}(k) - \mathbf{w}_j^T \mathbf{x}(k) \geq 1 - \xi_k, \\ &\quad\quad \forall i \in A\big(y(k)\big), \forall j \in S(i), \forall k \in [1, \cdots, N] \\ \textbf{where}: &\quad \xi_k \geq 0, \forall k \in [1, \cdots, N] \end{aligned} \tag{2.14}$$

where $\lambda > 0$ is a penalty parameter, ξ_k are the slack variables, and $\boldsymbol{K}_{ij} \geq 0$ captures the hierarchical structure relationships. More precisely, $\boldsymbol{K}_{ij} = 0$, if node i is neither an ancestor nor a descendant of node j; otherwise $\boldsymbol{K}_{ij} > 0$.

2.6.2 Shrinking Data-Sparse Leaf Node Model Parameters Toward Data-Rich Ancestor Nodes

Insufficient examples available for training at the leaf categories is one of the main reasons for inferior classification performance. To overcome this, a well-established statistical technique known as *shrinkage* is explored in the paper by McCallum et al. [27]. Learned model parameter estimates of the data-sparse leaf nodes are generalized by enforcing the parameter smoothness with the data-rich ancestor nodes. Probabilistically, the marginal probability of generating an input instance $\mathbf{x}(i)$ given the model parameters for leaf categories $\big[\mathbf{W}\big]_{L*d} = [\mathbf{w}_1^T, \cdots, \mathbf{w}_L^T]$ is given by the sum of total probability over individual components using the equation shown in Eq. (2.15):

$$P\big(\mathbf{x}(i)|\mathbf{W}\big) = \sum_{l \in L} P(l|\mathbf{W})P\big(\mathbf{x}(i)|l; \mathbf{W}\big) \tag{2.15}$$

where $P\big(\mathbf{x}(i)|l; \mathbf{W}\big)$ can be computed as the product of probability estimates for individual input component as shown in Eq. (2.16):

$$P\big(\mathbf{x}(i)|l; \mathbf{W}\big) = P(|\mathbf{x}(i)|) \prod_{k=1}^{|\mathbf{x}(i)|} P\big(\mathbf{x}_k(i)|l; \mathbf{W}\big) \tag{2.16}$$

Given the initial parameter estimates of the learned model weight vectors corresponding to class l and its ancestor as $\big\{\mathbf{w}_l^i\big\}_{i \in A(l)}$, parameter smoothing can

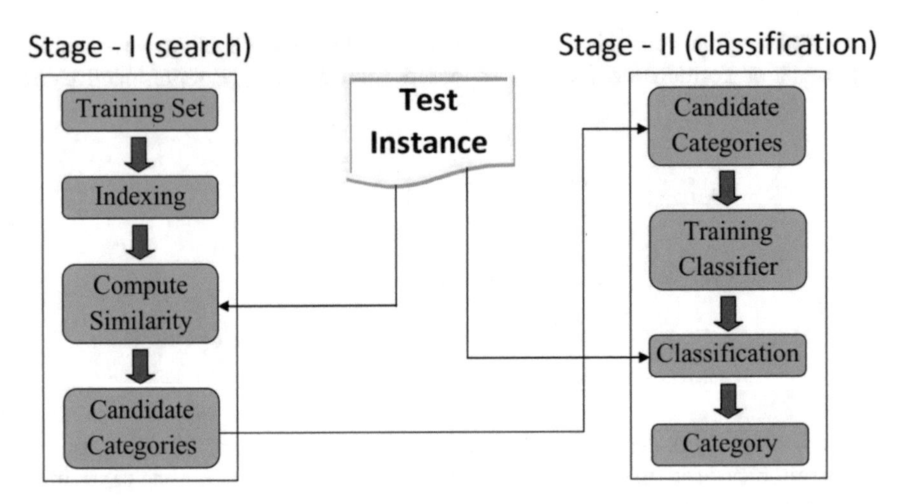

Fig. 2.10 Flowchart for two-stage classification

be applied by shrinking the learned parameters of class l to its ancestor categories as per the rule shown in Eq. (2.17):

$$\mathbf{w}_l = \sum_{i \in A(l)} \lambda_l^i \mathbf{w}_l^i \tag{2.17}$$

where λ_l^i denotes the shrinkage parameter.

2.6.3 *Two-Stage Classification for Large-Scale Taxonomy*

LSHC problems are characterized by the extremely large and deep hierarchy. As the hierarchy size increases, learning models for every node in the hierarchy becomes difficult task. To address this issue, Xue et al. [38] proposed two-stage classification (referred to as deep classification) approach as shown in Fig. 2.10. For each test document, in the first stage (also known as search stage), a set of candidate categories is retrieved based on similarity to the test document. Then the second stage (also known as classification stage) builds a classifier on the hierarchy restricted to the set of categories fetched in first stage (as shown in Fig. 2.11) and classifies the test document using the restricted hierarchy. Although pruning reduces the hierarchy to a manageable size, one severe drawback of this approach is having to train a different classifier for each test document. Some of the important observations pointed out in this paper is as follows:

1. Deep classification method is very effective at lower levels in the hierarchy (or deep hierarchy).

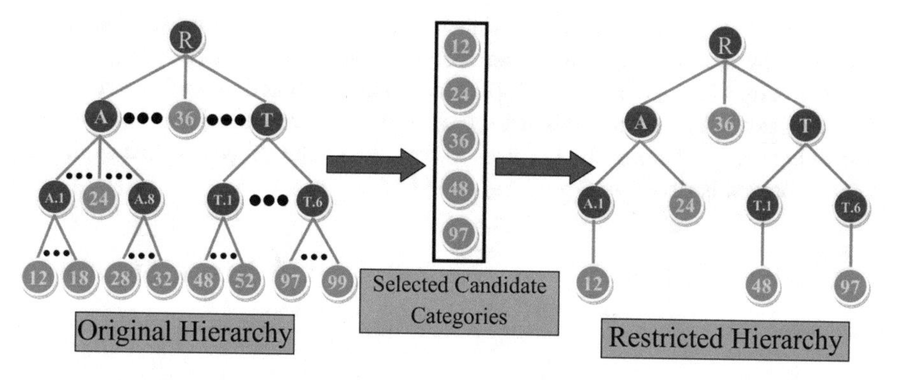

Fig. 2.11 Figure showing an example of restricted hierarchy that is generated from the set of selected candidate categories

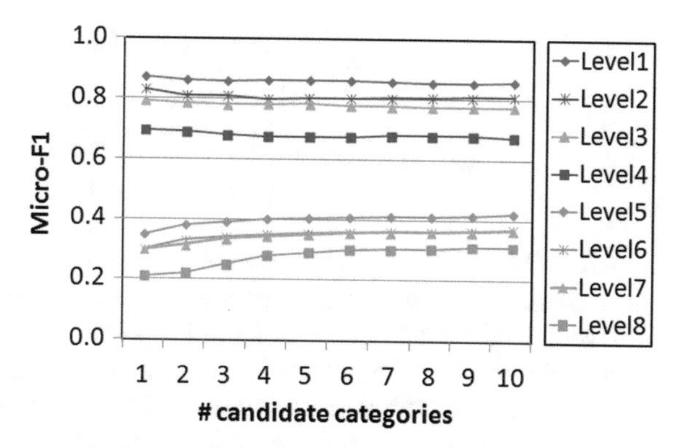

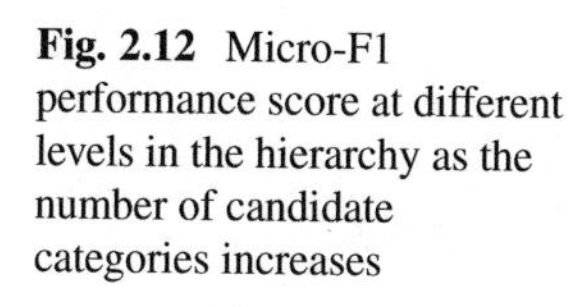

Fig. 2.12 Micro-F1 performance score at different levels in the hierarchy as the number of candidate categories increases

2. As the number of candidate categories chosen by the search stage increases, chances for finding the correct label for test example in the classification stage increase but with increase in evaluation time. Further, improvement is much more at lower levels in the hierarchy (deep hierarchy) as shown in Fig. 2.12.

2.6.4 Parent-Child Regularization

Traditional methods for classification learn classifiers for each leaf node (task) to discriminate one class from the other. This method works well *iff* dataset is small and well-balanced and there are sufficient positive examples per class to learn generalized discriminant function. However, these conditions are not applicable for large-scale problem as datasets are often unbalanced with lots of rare categories (more than 75% of LSHTC datasets belongs to rare categories as shown in Fig. 2.8). To deal with these issues, Gopal et al. proposed regularization framework [18] that incorporates interclass dependencies to improve classification.

In this approach, hierarchical dependencies between different classes are exploited by enforcing the model parameters (weights) of each class to be similar to their parent in regularization. Depending upon loss function used for classifiers training, two different methods were proposed, HR-SVM and HR-LR. HR-SVM uses hinge loss whereas HR-LR uses logistic loss. The proposed formulation of their approach is shown in Eqs. (2.18) and (2.19):

$$\textbf{HR-SVM} \quad \underset{\mathbf{W}}{\textbf{minimize}} \sum_{n \in \mathcal{N}} \frac{1}{2} ||\mathbf{w}_n - \mathbf{w}_{\pi(n)}||^2 + \lambda \sum_{l \in L} \sum_{i=1}^{N} \left|1 - y_l(i)\mathbf{w}_l^T \mathbf{x}(i)\right|_+ \tag{2.18}$$

$$\textbf{HR-LR} \quad \underset{\mathbf{W}}{\textbf{minimize}} \sum_{n \in \mathcal{N}} \frac{1}{2} ||\mathbf{w}_n - \mathbf{w}_{\pi(n)}||^2 + \lambda \sum_{l \in L} \sum_{i=1}^{N} \textbf{log}\Big(1 + exp\big(- y_l(i)\mathbf{w}_l^T \mathbf{x}(i)\big)\Big) \tag{2.19}$$

$$\textbf{where} : y_l(i) = \begin{cases} 1 & \text{if } \mathbf{x}(i) \text{ belongs to class } l \in L \\ -1 & \textit{otherwise} \end{cases}$$

HR-SVM and HR-LR methods are examples of global classification approach since all model parameters are learned together. It gives state-of-the-art performance results on LSHTC datasets. It can also be easily parallelized making it suitable for large-scale problem. Parallelization is achieved by optimizing the alternate even and odd levels at subsequent iterations. This is possible because each node is independent of all other nodes except its neighbor. Steps for parallel optimization are as follows:

1. Fix odd-level parameters, optimize even levels in parallel
2. Fix even-level parameters, optimize odd levels in parallel
3. Repeat until convergence

Proposed formulation can also be extended to graph by first finding the minimum graph coloring [Np-hard] and repeatedly optimizing nodes with the same color in parallel during each iteration.

To demonstrate the effectiveness of this approach, experimental results are provided below for HR-SVM and HR-LR. Figures 2.13 and 2.14 show the performance comparison of HR-SVM (and HR-LR) against equivalent flat baseline SVM (and LR) models. Clearly, HR-SVM (and HR-LR) outperforms SVM (and LR) for all datasets. Computationally, HR-SVM (and HR-LR) is expensive to train in comparison to SVM (and LR) because complex global optimization is involved in the training of HR-SVM (and HR-LR).

For completeness, Table 2.4 shows the performance comparison of HR-SVM and HR-LR approach against other HC approaches. Again, HR-SVM provides best results with few exceptions where HBLR is better. Although HBLR is superior

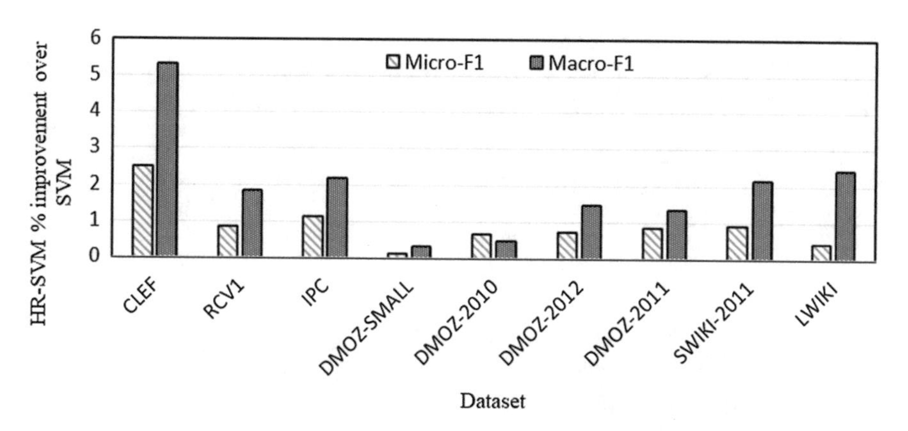

Fig. 2.13 Percentage performance improvement of HR-SVM over SVM for various datasets

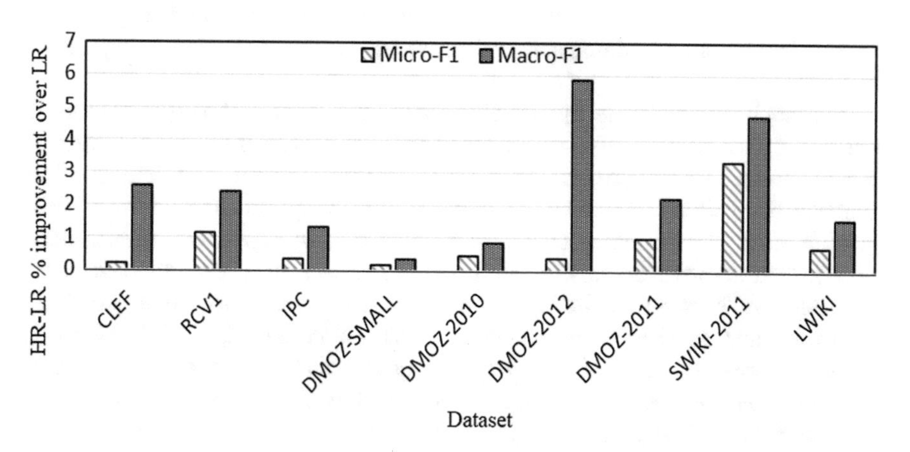

Fig. 2.14 Percentage performance improvement of HR-LR over LR for various datasets

for some datasets, it is computationally expensive and hence not scalable for large datasets. Also, HBLR is not applicable on multi-label datasets.

2.6.5 Cost-Sensitive Learning

Although recursive regularization has been shown to be effective in improving the HC performance, they induce large-scale optimization problems which require specialized solutions [7]. Even though distributed optimization methods are able to scale well to extremely large-scale scenarios, the global optimization of all the model parameters in an integrated fashion incurs communication overhead, which can sometimes be considerable. To address these shortcomings, we need a solution

Table 2.4 Micro-F1 performance comparison [18]

Datasets	HR-SVM	HR-LR	TD [26]	HSVM [36]	HOT [41]	HBLR [19]
CLEF	80.02	80.12	70.11	79.72	73.84	**81.41**
RCV1	**81.66**	81.23	71.34	NA	NS	NA
IPC	54.26	55.37	50.34	NS	NS	**56.02**
DMOZ-SMALL	45.31	45.11	38.48	39.66	37.12	**46.03**
DMOZ-2010	**46.02**	45.84	38.64	NS	NS	NS
DMOZ-2012	**57.17**	53.18	55.14	NS	NS	NS
DMOZ-2011	**43.73**	42.27	35.91	NA	NS	NA
SWIKI-2011	**41.79**	40.99	36.65	NA	NA	NA
LWIKI	**38.08**	37.67	NA	NA	NA	NA

NA not applicable, *NS* not scalable

that decouples models so that they can be trained in parallel while being effective in HC. Cost-sensitive learning approach provides alternative solution that addresses this issue.

Misclassifications made by the classifiers are not equal for all instances in the sense that nearby misclassifications are less severe than those farther in the hierarchy. This approach incorporates misclassification cost into the loss function while learning models. Cost assigned is based on the severity of misclassification. For example, *Dog* misclassified as *Cat* has lower misclassification cost in comparison to *Dog* misclassified as *Soccer*. Learning models for cost-sensitive learning is similar to flat method but with cost value associated with each instance in the loss function as shown in Eq. (2.20). This approach is also referred to as HierCost in the paper [8], and there are many variants depending on how the misclassification cost is computed. One of the commonly used methods for assigning misclassification cost is based on the hierarchical distance.

$$\Psi_n = \min_{\mathbf{w}_n} \left[\sum_{i=1}^{N} \omega_i^n \log\left(1 + \exp\left(-y_n(i)\mathbf{w}_n^T \mathbf{x}(i)\right)\right) + \frac{\lambda}{2} \|\mathbf{w}_n\|_2^2 \right] \tag{2.20}$$

where ω_n^i is the importance of example i for training a model at node n.

HierCost approach can be further extended to improve classification performance for the skewed category distribution in large-scale datasets where majority of the classes belong to rare categories. Imbalance cost which is defined using squashing function can be added to the loss function (in addition to misclassification cost) while optimization. Resultant effect of adding imbalance cost is that misclassification made for rare categories is severely penalized in comparison to classes with lots of examples, thereby avoiding decision boundaries favoring large classes. For illustrative purpose, Figs. 2.15 and 2.16 show the performance comparison of HierCost with flat LR model and other hierarchical models.

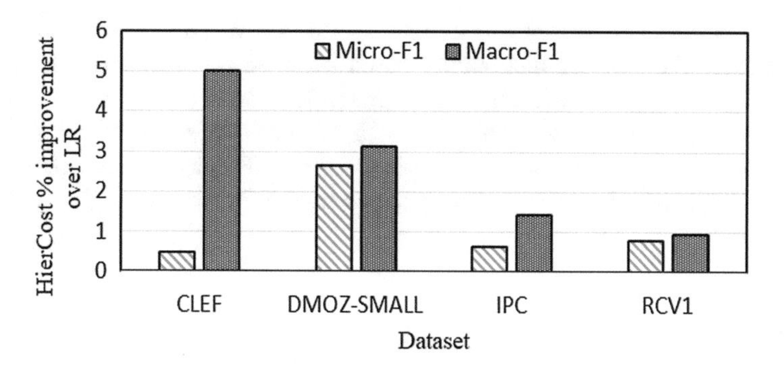

Fig. 2.15 Percentage performance improvement of HierCost over LR for various datasets

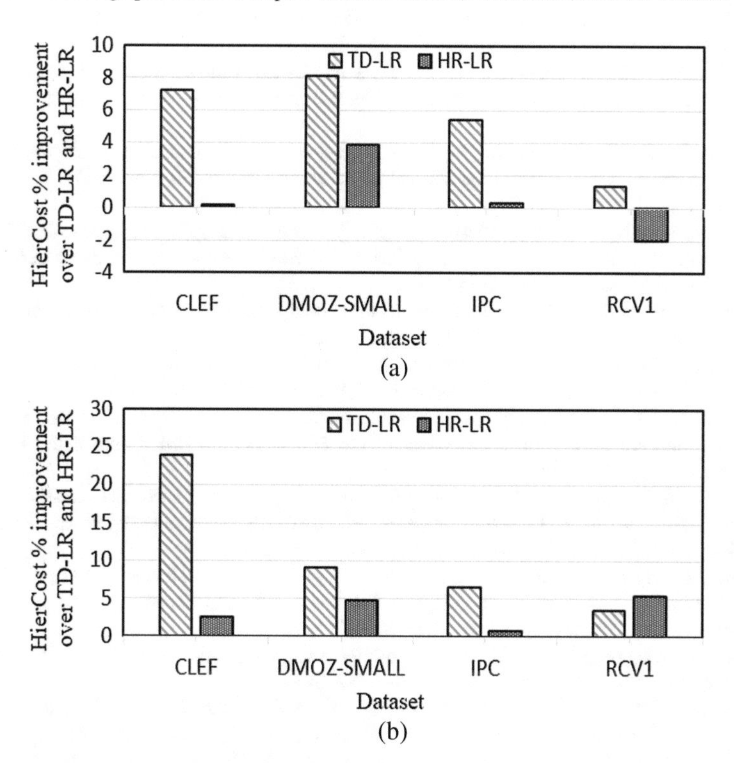

Fig. 2.16 (**a**) Micro-F1 and (**b**) Macro-F1 performance comparison of HierCost over TD-LR and HR-LR for various datasets

2.6.6 *Refined Experts*

Hierarchical classification problem suffers from two significant challenges—*error propagation* and increasingly complex *nonlinear decision surfaces* at higher levels in the hierarchy. To overcome this problem, Bennett et al. [4] proposed the method of refined experts, where refinement method is used to eliminate the error propagation by changing the training distribution based on cross-validation results. Further, expert extraction of meta-features at lower levels is done to improve the decision boundary at higher levels. Combining both of these steps into TD classification settings is referred by the author as *refined experts*. Empirical evaluation of their proposed approach shows an improvement up to 30% in F1 score.

References

1. Babbar, R., Maundet, K., Schölkopf, B.: Tersesvm: A scalable approach for learning compact models in large-scale classification. In: Proceedings of the 2016 SIAM International Conference on Data Mining, pp. 234–242 (2016)
2. Babbar, R., Partalas, I., Gaussier, E., Amini, M.R.: On flat versus hierarchical classification in large-scale taxonomies. In: Advances in Neural Information Processing Systems, pp. 1824–1832 (2013)
3. Babbar, R., Partalas, I., Gaussier, E., Amini, M.R., Amblard, C.: Learning taxonomy adaptation in large-scale classification. The Journal of Machine Learning Research **17**(1), 3350–3386 (2016)
4. Bennett, P.N., Nguyen, N.: Refined experts: improving classification in large taxonomies. In: Proceedings of the 32nd international ACM SIGIR conference on Research and development in information retrieval, pp. 11–18 (2009)
5. Ceci, M., Malerba, D.: Classifying web documents in a hierarchy of categories: a comprehensive study. Journal of Intelligent Information Systems **28**(1), 37–78 (2007)
6. Cesa-Bianchi, N., Gentile, C., Zaniboni, L.: Incremental algorithms for hierarchical classification. Journal of Machine Learning Research **7**(Jan), 31–54 (2006)
7. Charuvaka, A., Rangwala, H.: Approximate block coordinate descent for large scale hierarchical classification. In: Proceedings of the 30th Annual ACM Symposium on Applied Computing, pp. 837–844 (2015)
8. Charuvaka, A., Rangwala, H.: Hiercost: Improving large scale hierarchical classification with cost sensitive learning. In: Joint European Conference on Machine Learning and Knowledge Discovery in Databases (ECML/PKDD), pp. 675–690 (2015)
9. Costa, E., Lorena, A., Carvalho, A., Freitas, A.: Top-down hierarchical ensembles of classifiers for predicting g-protein-coupled-receptor functions. In: Advances in Bioinformatics and Computational Biology, pp. 35–46. Springer (2008)
10. Crammer, K., Singer, Y.: On the learnability and design of output codes for multiclass problems. Machine learning **47**(2–3), 201–233 (2002)
11. Dekel, O., Keshet, J., Singer, Y.: Large margin hierarchical classification. In: Proceedings of the twenty-first International Conference on Machine Learning (ICML), p. 27 (2004)
12. Dimitrovski, I., Kocev, D., Loskovska, S., Džeroski, S.: Hierarchical annotation of medical images. Pattern Recognition **44**(10), 2436–2449 (2011)
13. Dimitrovski, I., Kocev, D., Loskovska, S., Džeroski, S.: Hierarchical classification of diatom images using predictive clustering trees. Ecological Informatics **7**, 19–29 (2012)
14. Dumais, S., Chen, H.: Hierarchical classification of web content. In: Proceedings of the 23rd annual International ACM SIGIR Conference on Research and Development in Information Retrieval, pp. 256–263 (2000)
15. Eisner, R., Poulin, B., Szafron, D., Lu, P., Greiner, R.: Improving protein function prediction using the hierarchical structure of the gene ontology. In: Proceedings of the 2005 IEEE Symposium on Computational Intelligence in Bioinformatics and Computational Biology (CIBCB), pp. 1–10 (2005)

16. Fagni, T., Sebastiani, F.: On the selection of negative examples for hierarchical text categorization. In: Proceedings of the 3rd Language and Technology Conference (2007)
17. Gopal, S., Yang, Y.: Distributed training of large-scale logistic models. In: Proceedings of the 30th International Conference on Machine Learning (ICML), pp. 289–297 (2013)
18. Gopal, S., Yang, Y.: Recursive regularization for large-scale classification with hierarchical and graphical dependencies. In: Proceedings of the 19th ACM SIGKDD International Conference on Knowledge Discovery and Data mining, pp. 257–265 (2013)
19. Gopal, S., Yang, Y., Bai, B., Niculescu-Mizil, A.: Bayesian models for large-scale hierarchical classification. In: Advances in Neural Information Processing Systems, pp. 2411–2419 (2012)
20. Holden, N., Freitas, A.A.: A hybrid particle swarm/ant colony algorithm for the classification of hierarchical biological data. In: SIS, pp. 100–107 (2005)
21. Hsieh, C.J., Chang, K.W., Lin, C.J., Keerthi, S.S., Sundararajan, S.: A dual coordinate descent method for large-scale linear SVM. In: Proceedings of the 25th International Conference on Machine Learning (ICML), pp. 408–415 (2008)
22. Koller, D., Sahami, M.: Hierarchically classifying documents using very few words. In: Proceedings of the Fourteenth International Conference on Machine Learning (ICML), pp. 170–178 (1997)
23. Kosmopoulos, A., Partalas, I., Gaussier, E., Paliouras, G., Androutsopoulos, I.: Evaluation measures for hierarchical classification: a unified view and novel approaches. Data Mining and Knowledge Discovery **29**(3), 820–865
24. Lewis, D.D., Yang, Y., Rose, T.G., Li, F.: Rcv1: A new benchmark collection for text categorization research. Journal of machine learning research **5**(Apr), 361–397 (2004)
25. Li, T., Ogihara, M.: Music genre classification with taxonomy. In: IEEE International Conference on Acoustics, Speech, and Signal Processing, vol. 5, pp. v–197 (2005)
26. Liu, T.Y., Yang, Y., Wan, H., Zeng, H.J., Chen, Z., Ma, W.Y.: Support vector machines classification with a very large-scale taxonomy. ACM SIGKDD Explorations Newsletter **7**(1), 36–43 (2005)
27. McCallum, A., Rosenfeld, R., Mitchell, T.M., Ng, A.Y.: Improving text classification by shrinkage in a hierarchy of classes. In: Proceedings of the 15th International Conference on Machine Learning (ICML), vol. 98, pp. 359–367 (1998)
28. Naik, A., Rangwala, H.: A ranking-based approach for hierarchical classification. In: IEEE International Conference on Data Science and Advanced Analytics (DSAA), pp. 1–10 (2015)
29. Naik, A., Rangwala, H.: Embedding feature selection for large-scale hierarchical classification. In: Proceedings of the IEEE International Conference on Big Data, pp. 1212–1221 (2016)
30. Naik, A., Rangwala, H.: Hierflat: flattened hierarchies for improving top-down hierarchical classification. International Journal of Data Science and Analytics **4**(3), 191–208 (2017)
31. Naik, A., Rangwala, H.: Improving large-scale hierarchical classification by rewiring: A data-driven filter based approach. Journal of Intelligent Information Systems (JIIS) pp. 1–24 (2018)
32. Paes, B.C., Plastino, A., Freitas, A.A.: Improving local per level hierarchical classification. Journal of Information and Data Management **3**(3), 394 (2012)
33. Secker, A., Davies, M.N., Freitas, A.A., Timmis, J., Mendao, M., Flower, D.R.: An experimental comparison of classification algorithms for the hierarchical prediction of protein function. Expert Update (the BCS-SGAI Magazine) **9**(3), 17–22
34. Silla Jr, C.N., Freitas, A.A.: A survey of hierarchical classification across different application domains. Data Mining and Knowledge Discovery **22**(1–2), 31–72 (2011)
35. Sokolova, M., Lapalme, G.: A systematic analysis of performance measures for classification tasks. Information Processing and Management **45**(4), 427–437 (2009)
36. Tsochantaridis, I., Joachims, T., Hofmann, T., Altun, Y.: Large margin methods for structured and interdependent output variables. Journal of machine learning research **6**(Sep), 1453–1484 (2005)
37. Vens, C., Struyf, J., Schietgat, L., Džeroski, S., Blockeel, H.: Decision trees for hierarchical multi-label classification. Machine learning **73**(2), 185 (2008)

38. Xue, G.R., Xing, D., Yang, Q., Yu, Y.: Deep classification in large-scale text hierarchies. In: Proceedings of the 31st annual International ACM SIGIR Conference on Research and Development in Information Retrieval, pp. 619–626 (2008)
39. Yang, Y.: An evaluation of statistical approaches to text categorization. Information retrieval **1**(1–2), 69–90 (1999)
40. Yang, Y.: A study of thresholding strategies for text categorization. In: Proceedings of the 24th annual international ACM SIGIR conference on Research and development in information retrieval, pp. 137–145 (2001)
41. Zhou, D., Xiao, L., Wu, M.: Hierarchical classification via orthogonal transfer. In: Proceedings of the 28th International Conference on Machine Learning (ICML), pp. 801–808 (2011)
42. Zhu, S., Wei, X.Y., Ngo, C.W.: Error recovered hierarchical classification. In: Proceedings of the 21st ACM international conference on Multimedia, pp. 697–700 (2013)
43. Zimek, A., Buchwald, F., Frank, E., Kramer, S.: A study of hierarchical and flat classification of proteins. IEEE/ACM Transactions on Computational Biology and Bioinformatics **7**(3), 563–571 (2010)

Chapter 3
Hierarchical Structure Inconsistencies

Hierarchies are useful for improving classification performance. It provides useful structural relationships among different classes that can be exploited for learning generalized classification models. In the past, researchers have demonstrated the usefulness of hierarchies for classification and have obtained promising results [2, 4, 6, 9, 16]. Utilizing the hierarchical structure has also been shown to improve the classification performance for rare categories (having ≤ 10 examples) as well [8].

Top-down HC methods that leverage the hierarchy during the learning and prediction process are effective approaches to deal with large-scale problems [6]. Classification decision for top-down methods involves invoking only the models in the relevant path within the hierarchy. Though computationally efficient, these methods have higher number of misclassifications due to error propagation, [10] i.e., error made at higher levels cannot be corrected at the lower levels. In many situations, hierarchies used for learning models are not consistent due to the presence of *inconsistent* nodes and links resulting in excessive error propagation. As a result, HC approaches are outperformed by the flat classifiers that completely ignore the hierarchy [19, 22]. Figure 3.1 shows an example of inconsistent nodes and links at node A. Learning generalized classifier at node A becomes difficult due to heterogeneity of classes which in turn results in many misclassifications at this node.

Definition 3.1 Inconsistent node—Nodes in the hierarchy that contain examples from other branches in the hierarchy are known as inconsistent nodes. In general, inconsistent node is a heterogeneous mixture of examples from different classes making it difficult to learn generalized classifier. Node A in Fig. 3.1 is inconsistent because it contains categories $A.2$ and $A.3$ which should belong to nodes C and B, respectively, because of similarities between examples.

A. Naik, H. Rangwala, *Large Scale Hierarchical Classification: State of the Art*,
SpringerBriefs in Computer Science, https://doi.org/10.1007/978-3-030-01620-3_3

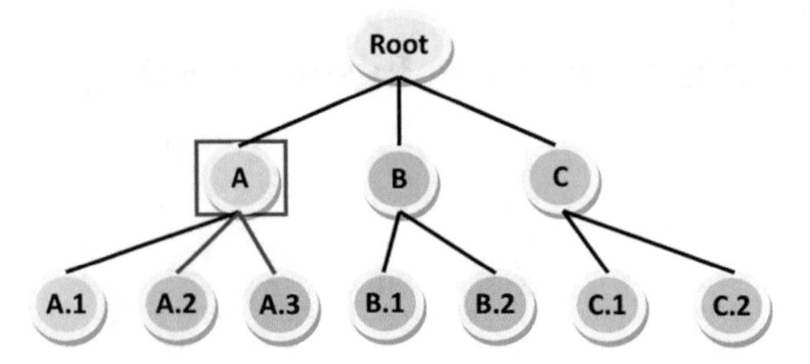

Fig. 3.1 [Best viewed in color] Hierarchy with inconsistent node and links (marked in red). Similar classes are shown using the same color

Definition 3.2 Inconsistent link—Parent-child link in the hierarchy that makes parent node inconsistent is known as inconsistent link. In Fig. 3.1 links $A - A.2$ and $A - A.3$ are inconsistent because $A.2$ and $A.3$ shouldn't be the children of parent node A.

3.1 Hierarchical Restructuring Experiment

Flat classification works well for well-balanced datasets with smaller number of categories, but it has expensive train/prediction cost. On the other hand, HC is computationally efficient and is preferable for large-scale datasets. It also performs well for rare categories by leveraging hierarchical structure. However, for some benchmark datasets, flat methods (and its variant) have a good performance in comparison to hierarchical methods. This observation provides motivation to look into alternative perspective for solving HC problem rather than learning complex models. To this end, deeper analysis was performed in identifying the impact of hierarchical structure on HC performance. Basically, the following questions are critical to answer before we explore further on modifying hierarchical structure to improve classification performance:

1. Can predefined experts hierarchy always be trusted to achieve good classification performance?
2. Can hierarchy be tweaked (or adjusted) to improve the performance?

To investigate, case study was done on pruned hierarchy of Newsgroup dataset.[1] Figure 3.2a shows the expert-defined hierarchy, and Fig. 3.2b shows the adjusted

[1] http://qwone.com/~jason/20Newsgroups.

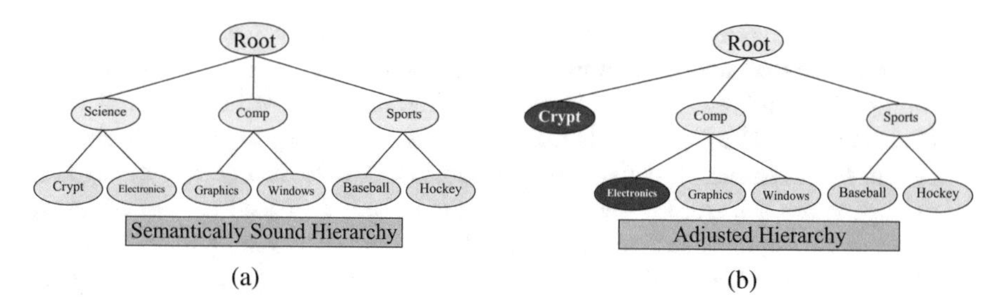

Fig. 3.2 Different hierarchies generated for subset of newsgroup dataset. (**a**) Semantically defined experts hierarchy and (**b**) adjusted hierarchy

hierarchy where *Science* node is deleted and its children *Crypt* and *Electronics* are linked to *Root* and *Comp* node, respectively. HC performance results (trained using LR models) showed that adjusted hierarchy has comparatively better performance in comparison to expert-defined hierarchy. This analysis provided insights that hierarchy can be modified to improve upon HC performance.

Data-driven hierarchy is more desirable over predefined experts hierarchy for achieving better HC performance.

3.2 Reasons for Hierarchical Inconsistencies

Most of the HC methods often use hierarchy during the learning process to design appropriate loss function for classification. Their performance can be severely affected if the hierarchy used for learning is not well-suited for classification purpose. In majority of the cases, the hierarchy used for training is manually designed by the domain experts that reflects the human view of the domain. This manual process of hierarchy creation suffers from various design issues and introduces inconsistencies that makes it unsuitable to achieve high classification accuracy. Some of the major reasons behind hierarchical inconsistencies includes:

1. Hierarchy is designed for the sole purpose of easy search and navigation without taking classification into consideration.
2. Hierarchical groupings of categories is done based on semantics, whereas classification depends on data characteristics such as *term frequency*.
3. A priori it is not clear to domain experts when to generate new nodes (i.e., hierarchy expansion) or merge two or more nodes to common parents (i.e., link creation) in the hierarchy, and it is often left at the discretion of domain experts to decide which results in certain degree of arbitrariness.

4. Consistent hierarchy design for datasets with large number of categories is prone to errors due to many confounding (confusing) classes.
5. Multiple hierarchies are possible for the same dataset based on experts' view. However, there is no consensus regarding which hierarchy is better for classification. For example, in categorizing products as shown in Fig. 3.3, the experts may generate a hierarchy by first separating products based on the company name (e.g., Apple, Microsoft) and then the product type (e.g., phone, tablet) or vice versa. Both hierarchies are equally good from the perspective of an expert. However, these different hierarchies may lead to different classification results.
6. Dynamic changes can affect hierarchical relationships over time, and therefore modifications within the hierarchy are needed as time passes. For example, as shown in Fig. 3.4, *Flood* is subgroup of *Geography* class, but when flood occurred in Chennai (India) in the year 2015, it becomes political news. Therefore, intuitively it makes more sense (and hence more beneficial for HC task) to make *Flood* as subgroup of *News* during that time.

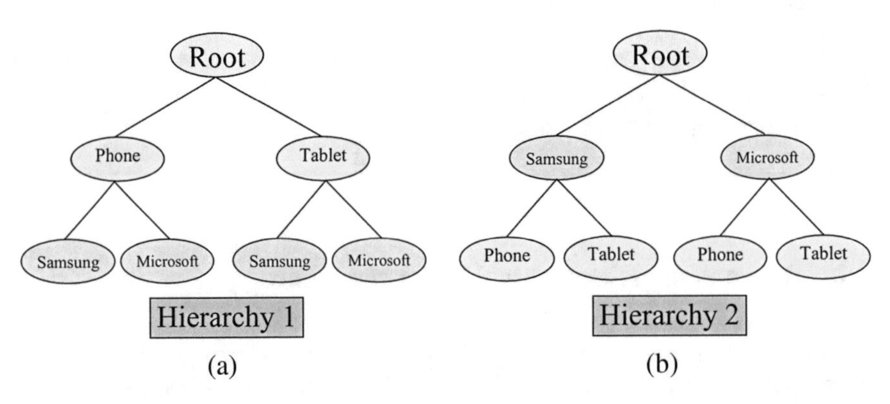

Fig. 3.3 Two different hypothetical hierarchies (**a**) and (**b**) possible from the same set of categories (Samsung, Microsoft, Phone, Tablet)

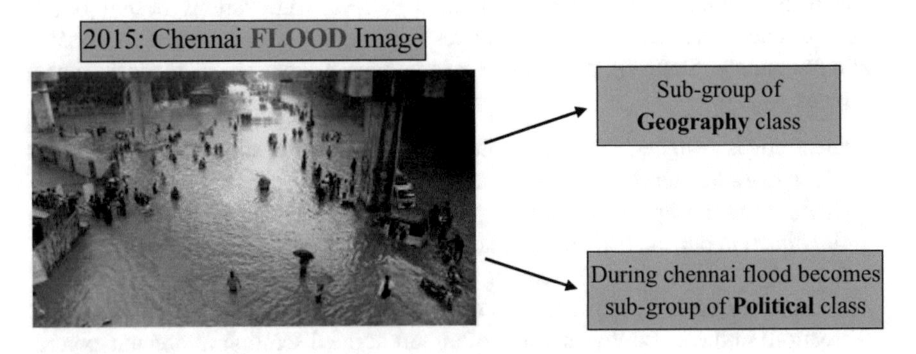

Fig. 3.4 Dynamic changes can affect hierarchical relationships

3.3 Different Methods for Hierarchy Restructuring

Several approaches that restructure the hierarchy have been developed in the past. Figure 3.5 provides the snapshot of different existing approaches in the literature. In this section, we will discuss in detail some of the most popular methods. Broadly, all methods can be divided into three categories:

1. Flattening approach
2. Rewiring approach
3. Clustering approach

3.3.1 *Flattening Approach*

In this approach, inconsistent nodes in the hierarchy are determined and flattened (removed). Based on different approaches followed for determining inconsistent nodes, it is further subcategorized:

1. Level flattening [18]—It is one of the approaches used in earlier works of hierarchy modification, where some of the levels are flattened (removed) from the original hierarchy prior to learning models. Based on levels flattened, various methods of hierarchy modification exist:
   (a) Top-Level Flattening (TLF)—As shown in Fig. 3.6b, TLF modifies the hierarchy by flattening the top level in the original hierarchy. Model learning and prediction for flattened hierarchy is done in similar fashion as TD methods.
   (b) Bottom-Level Flattening (BLF)—Similar to TLF with only difference, bottom level in the hierarchy is flattened instead of top level as shown in Fig. 3.6c.

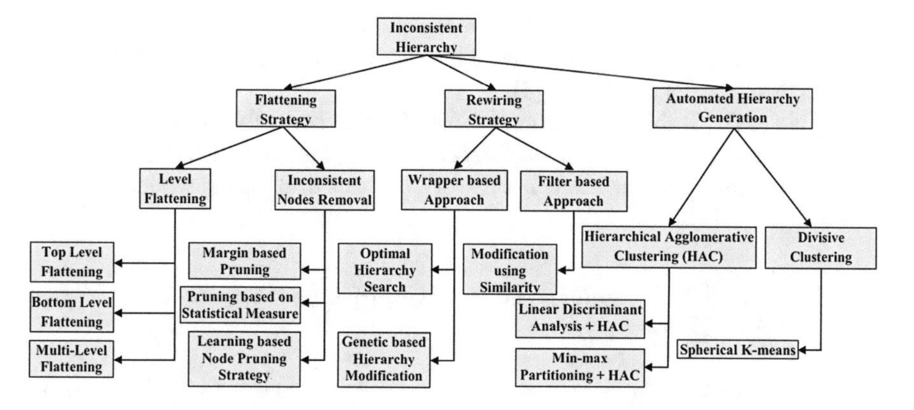

Fig. 3.5 Different approaches for solving hierarchical inconsistencies problem

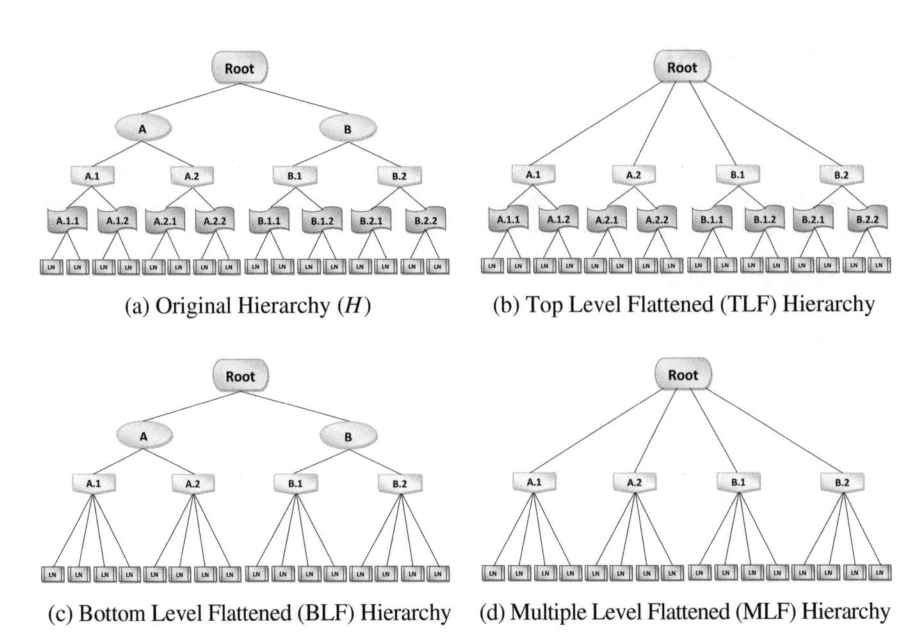

(a) Original Hierarchy (H) (b) Top Level Flattened (TLF) Hierarchy

(c) Bottom Level Flattened (BLF) Hierarchy (d) Multiple Level Flattened (MLF) Hierarchy

Fig. 3.6 Various hierarchical structures (**b**)–(**d**) obtained after flattening some of the levels from the original hierarchy shown in (**a**). "LN" denotes the leaf node in the figure

(c) Multiple-Level Flattening—Multiple levels are flattened in the hierarchy prior to model learning as shown in Fig. 3.6d.

2. Inconsistent node flattening (INF)—Level flattening approach although useful up to certain extent suffers from one major drawback—All nodes in the level are identified as inconsistent and flattened which may not be true, resulting in suboptimal classification performance. To overcome this, it is more reasonable to selectively remove some of the inconsistent nodes from the hierarchy based on certain measures such as SVM margins. Hierarchy modification using this approach is shown in Fig. 3.7b, and it is more beneficial for classification and has been theoretically justified in [5].

   Gao et al. [5] showed that for any classifier that correctly classifies m random input-output pairs using a set of $\mathscr{D}$ decision nodes, the generalization error bound with probability estimates greater than 1 - ζ is less than the expression shown in Eq. (3.1):

$$\frac{\delta r^2}{m}\left[\sum_{n\in\mathscr{D}}(\frac{1}{\gamma_n^2})\log(4em)\log(4m) + |\mathscr{D}|\log(2m) - \log(\frac{2}{\zeta})\right] \tag{3.1}$$

   where γ_n denotes the margin at node $n \in \mathscr{D}$, δ is a constant term, and r is the radius of the ball containing the distribution's support.

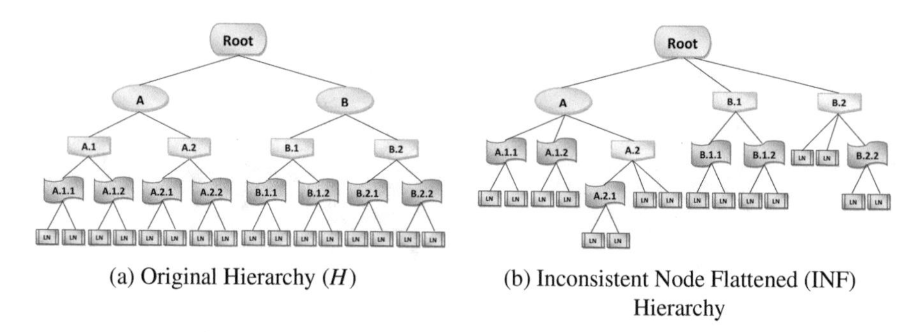

(a) Original Hierarchy (H)

(b) Inconsistent Node Flattened (INF) Hierarchy

Fig. 3.7 Hierarchical structure (**b**) obtained after flattening some of the inconsistent nodes from the original hierarchy shown in (**a**). "LN" denotes the leaf node in the figure

This provides two significant strategies in designing approach to reduce the generalization error: (a) Increase the margin γ_n for learned models at node $n \in \mathscr{N}$ in the hierarchy, or (b) Decrease the number of decision nodes $|\mathscr{D}|$ involved in making the prediction. For achieving the optimum classification performance, we need to balance the trade-off between the margin γ_n and the number of decision nodes $|\mathscr{D}|$. Two of the extreme cases for learning hierarchical classifiers are flat and top-down methods. For flat classifiers, we have to make single decision (i.e., $|\mathscr{D}| = 1$), but margin width γ_n is presumably small due to the large number of leaf categories that needs to be distinguished, which makes it difficult to obtain large margin. For top-down hierarchical classifiers, we have to make a series of decisions from root to leaf nodes (i.e., $|\mathscr{D}| \geq 1$) but margin γ_n is larger due to the fewer number of categories that needs to be distinguished at each of the decision nodes. Motivated by this trade-off, flattening-based method removes some of the inconsistent nodes in the hierarchy H, thereby increasing the value of margin γ_n for learned models at node n in the hierarchy, while minimizing the number of decision nodes to classify an unlabeled test instance.

Data-driven approach for removing inconsistent nodes in the expert-defined (original) hierarchy can lead to a generalized hierarchy that achieves higher classification performance irrespective of the hierarchical classifiers trained. In INF approach inconsistent nodes are selectively removed from the hierarchy. The criterion for flattening a node is based on certain measures such as SVM margins obtained for a node or optimal regularized risk minimization objective value attained by the model trained for that node on a separate validation set or degree of error made at the node. If the node n is identified as inconsistent, then it is flattened, i.e., remove n from the hierarchy and add its children to n's parent node. Based on the strategy adapted for identifying inconsistent nodes, many different approaches for INF hierarchy modification exist.

(a) Maximum-margin-based INF—Hierarchy modification using maximum-margin-based approach is proposed by Babbar et al. [1]. In this approach,

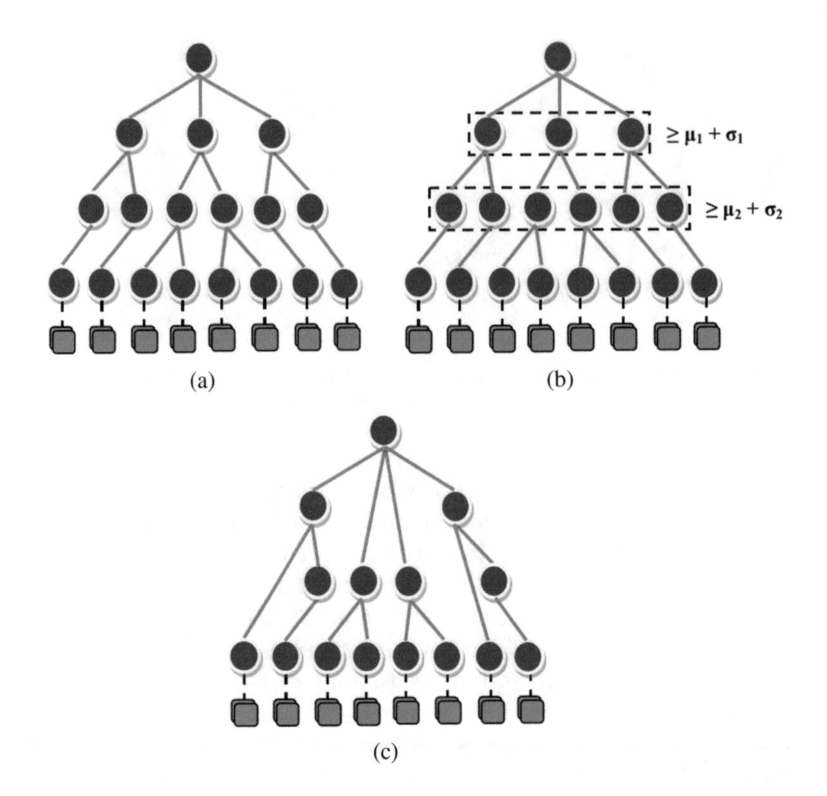

Fig. 3.8 Modified hierarchy obtained after removing inconsistent nodes from the hierarchy using Level-INF approach. (**a**) Original hierarchy (H). (**b**) Local flattening (Level-INF). (**c**) Modified hierarchy

inconsistent nodes are selectively removed from the hierarchy based on SVM margins rather than removing complete levels. Modified hierarchy obtained is then used for HC.

(b) Local approach for INF [10]—In this approach, inconsistent set of nodes are determined for each level based on objective value Ψ_n obtained for nodes n at that level. Criterion for flattening nodes is based on different defined thresholds for each level in the hierarchy. Statistical measures—mean (μ) and standard deviation (σ)—are used to define optimal threshold. Any node in the level with objective value greater than mean and standard deviation of that level is considered as inconsistent and is removed from the hierarchy. This approach is referred to as Level-INF in the paper. Figure 3.8 shows the original and modified hierarchy obtained after removing inconsistent nodes using Level-INF approach.

(c) Global approach for INF [10]—Different from Level-INF approach, global threshold is defined for the hierarchy instead of each level. Global threshold is computed by taking mean and standard deviation of objective values of all internal nodes in the hierarchy. Any nodes with objective value

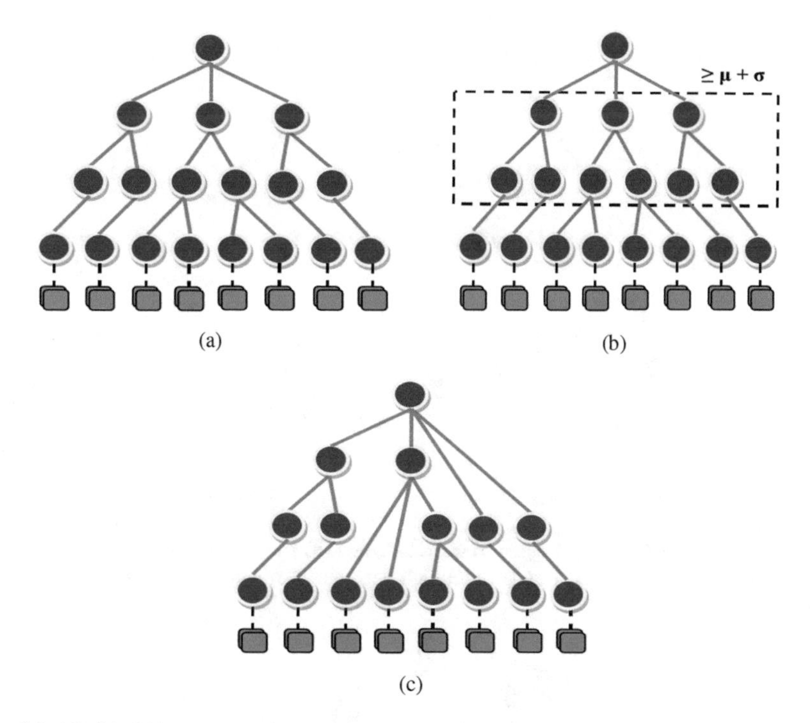

Fig. 3.9 Modified hierarchy obtained after removing inconsistent nodes from the hierarchy using Global-INF approach. (**a**) Original hierarchy (H). (**b**) Global flattening (Global-INF). (**c**) Modified hierarchy

greater than global threshold are identified as inconsistent and removed from the hierarchy. This approach is referred to as Global-INF in the paper. Figure 3.9 shows the original and modified hierarchy obtained after removing inconsistent nodes using Global-INF approach.

3.3.2 Rewiring Approach

Although flattening approach is useful for improving HC performance, it suffers from one limitation. It cannot deal with inconsistencies that exist in different parts of the hierarchy. As shown in Fig. 3.10c, rewiring approaches can resolve inconsistencies that exist in different parts of hierarchy. Different methods exist in the literature for performing rewiring in the original hierarchy. In this section, we will discuss some of the well-known methods.

1. Optimal hierarchy search [17]—This is a wrapper-based method that iteratively modifies the hierarchy by making one or few changes, which are then evaluated

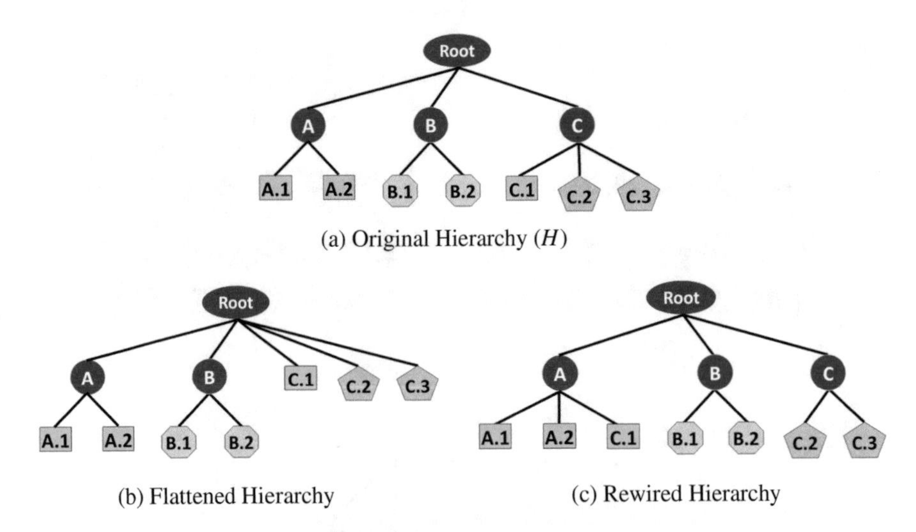

Fig. 3.10 Modified hierarchies obtained after flattening (**b**) and rewiring (**c**) approach applied on the original hierarchy (**a**). Leaf nodes with high degree of similarities are shown with the same color and shape

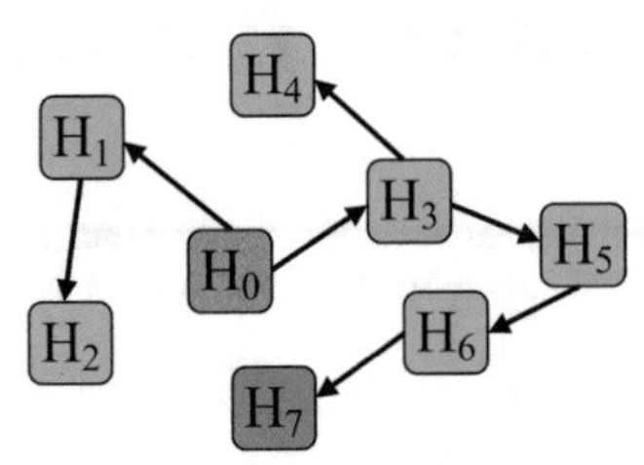

Fig. 3.11 Optimal hierarchy search in hierarchical space. H0 is the predefined (expert defined or original) hierarchy and H7 is the optimal hierarchy. H1–H6 are the intermediate hierarchies generated in the process of searching for optimal hierarchy

(on a validation set) by training a classification model to identify if the modified hierarchy has improved performance. Modified changes are retained if the performance results improve; otherwise the changes are discarded and the process is repeated. This repeated procedure of hierarchy modification continues until the optimal hierarchy that satisfies certain threshold criterion is reached (Fig. 3.11). Hierarchy modification at each step is performed using three defined elementary operations—promote, demote, and merge as shown in Fig. 3.12. This approach works on the assumption that optimal hierarchy is near the neighborhood of predefined hierarchy.

2. Genetic algorithm for improving performance—Considering only the current best obtained hierarchy for improving the performance may not be an optimal approach; therefore Qi and Davison [14] proposed using multiple best performing hierarchies at each step for improving the performance using genetic-based algorithm. Author proposed different methods for adapting the genetic operations

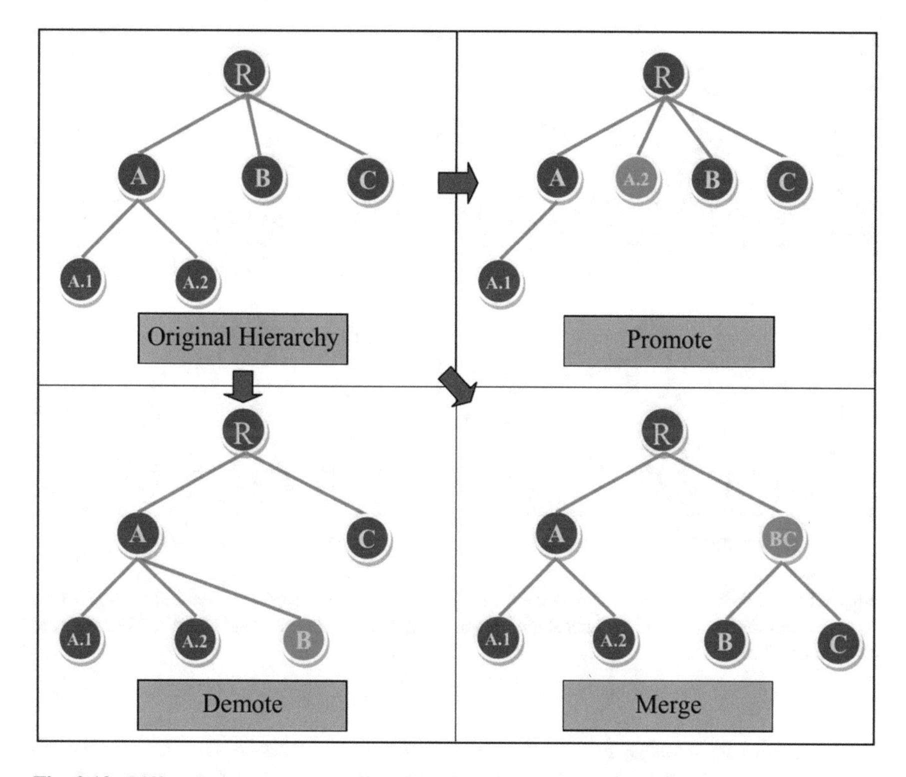

Fig. 3.12 Different elementary operations possible in the hierarchy

such as mutations and crossover operations to the hierarchical settings. Experiments on multiple classification tasks showed that their proposed algorithm can significantly improve classification task. However, the performance is highly dependent on the hierarchies and the operators selected at each step.

3. Filter-based rewiring approach—One of the major drawbacks of wrapper-based approaches [12, 14, 17] is that multiple hierarchies need to be evaluated prior to reaching the optimal hierarchy. This is an expensive process for large-scale datasets. To overcome this, filter based rewiring approach (*rewHier*) is proposed in the paper by Naik et al. [11]. They proposed three basic elementary operations—node creation, parent-child rewiring, and node deletion as shown in Fig. 3.13.

   Their approach consists of two major steps:

   (a) **Grouping Similar Class Pairs**—To ensure classes with high degree of similarity are grouped together under the same parent node in the modified taxonomy, this step identifies the similar class pairs that exist within the expert-defined hierarchy. Pairwise cosine similarity is used as the similarity measure as it is less prone to the curse of dimensionality [15]. Moreover, for computing similarities between two classes (let's say, class A and class B),

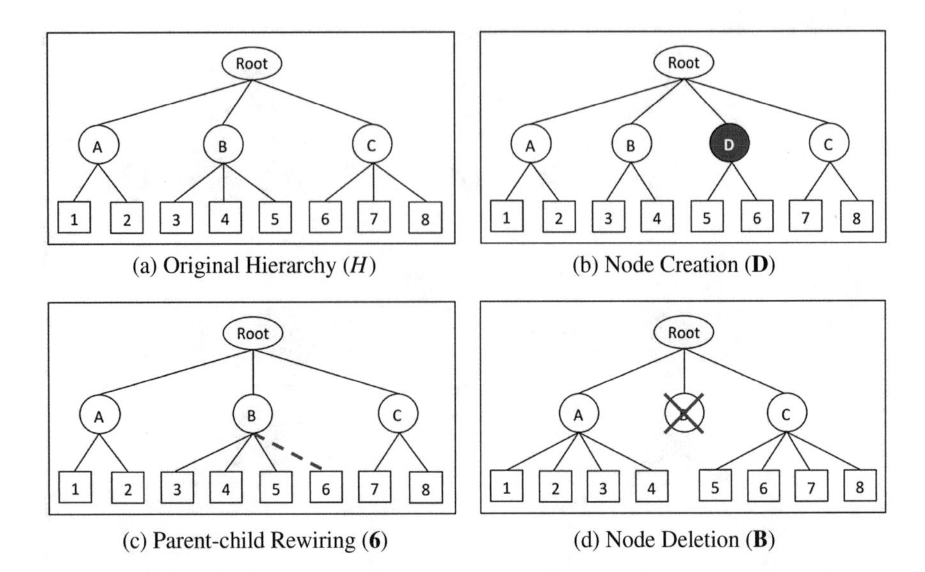

Fig. 3.13 Modified hierarchical structures (**b**)–(**d**) obtained after applying elementary operations to expert defined hierarchy (**a**). Leaf nodes are marked with "rectangle" and structural changes are shown in red color

similarity score is first computed between each instance in class A and each instance in class B and then averaged to get the final similarity score. Once the similarity scores are computed, set of similar pairs of classes **S** are determined using an empirically defined cutoff threshold τ that is determined from the dataset. For example, in Fig. 3.14a this step will group together the class pairs with high similarity scores such as $\mathbf{S} = \big[(religion.misc$ and $soc.religion.christian)$, $(electronics$ and $windows.x)$, $(electronics$ and $graphics), \cdots\big]$.

Pairwise similarity computation between different classes is one of the major bottleneck of this step. To make it scalable, similarity computation is distributed across multiple compute nodes. Given $\mathscr{L}$ number of classes, the total number of pairwise similarities that needs to be computed is given by Eq. (3.2):

$$^{L}C_2 = \frac{L * (L-1)}{2} \tag{3.2}$$

(b) Inconsistency Identification and Correction—To obtain the consistent hierarchy, most similar class pairs are grouped together to a common parent node in this step. Iteratively, starting from the most similar class pairs, potential inconsistencies (if the pairs of similar classes are in different branches (sub-trees)) are identified and corrected using elementary operation.

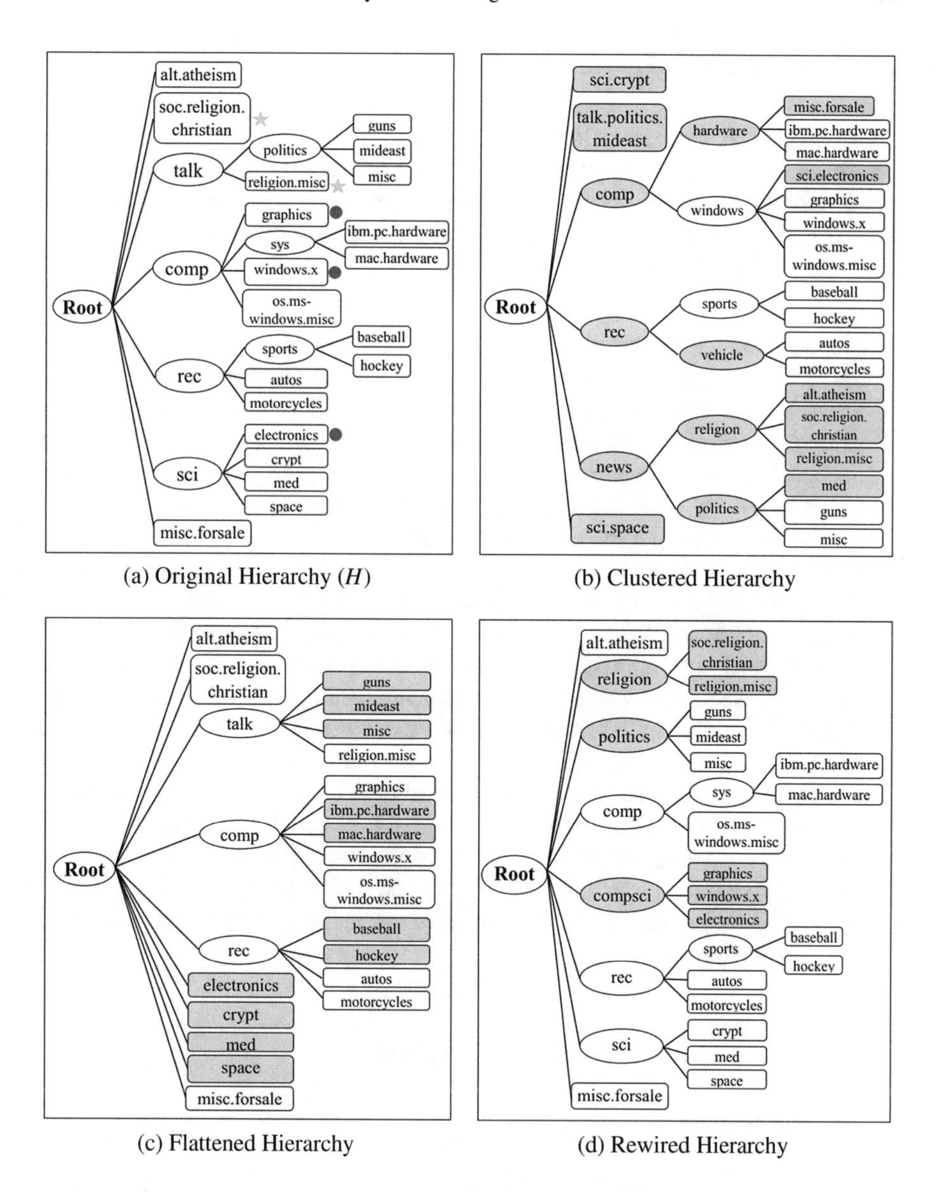

(a) Original Hierarchy (H)

(b) Clustered Hierarchy

(c) Flattened Hierarchy

(d) Rewired Hierarchy

Fig. 3.14 [Best viewed in color] NewsGroup dataset: (**a**) Expert-defined hierarchy (classes with high degree of similarities are marked with the same symbols, i.e., circle, star) modified using various methods. (**b**) Agglomerative clustering with cluster cohesion to restrict the height to original height. (**c**) Global-INF flattening method. (**d**) Filter-based rewiring (*rewHier*) method. Modified structural changes in comparison to baseline expert-defined hierarchy are shown in green color

3.3.3 Clustering Approach

Clustering-based approaches have also been adapted in some of the studies where consistent hierarchy is generated from scratch using agglomerative or divisive clustering algorithms [7, 13, 21]. In this approach, classes are grouped based on their overall similarity to one another. We will discuss two different approaches for clustering that generates hierarchy for performing HC:

1. Agglomerative clustering with linear discriminant projection—Li et al. [7] proposed the use of linear discriminant projection to transform all instances to lower dimensional space before performing the hierarchical agglomerative clustering, which produces meaningful hierarchy of clusters. Hierarchies generated using this approach showed improved classification performance, but the main drawback of their approach is that they ignore the original hierarchical structure which may carry some important information. In addition, there is no theoretical guarantee for deciding the number of levels that can be used to achieve the best classification performance. Moreover, this approach is practically not suitable for large-scale problems due to the projection step which is not scalable.
2. Divisive clustering with Fisher index criteria—Punera et al. [13] proposed divisive clustering approach to create hierarchy from the set of predefined classes. In their approach, first discriminant features are selected using the Fisher index criteria. Classes are then recursively divided into two clusters using spherical K-means until all the leaf nodes have instances from only one class. During the division process, there is possibility of class being split into multiple classes when it cannot be confidently assigned to either of the two newly created clusters (i.e., more than certain fraction of the instances from class cannot be assigned to either cluster). Again, the problem with this approach is scalability for large-scale problems, and original hierarchy is ignored which might contain vital information.

3.4 Experimental Results and Analysis

3.4.1 Case Study

To understand the qualitative difference between hierarchy generated using various approaches discussed in previous section, case study was performed on newsgroup dataset. The dataset has 11,269 training instances, 7505 test instances, 20 classes, and 3 levels. Different hierarchical structures (expert defined, clustered, flattened, and rewired) obtained after applying hierarchy modification are shown in Fig. 3.14d. To understand the importance of each hierarchy for classification, first top-down HC model is learned on this hierarchy separately, and then learned model is evaluated on test dataset. This step is repeated in each of the hierarchy by randomly selecting five different sets of training and test split in the same ratio as original dataset.

Table 3.1 μF_1 and MF_1 performance comparison using different hierarchy modification approaches on newsgroup dataset

Metric	TD-LR [Fig. 3.14a]	Clustering [7] (agglomerative) [Fig. 3.14b]	Flattening [10] (Global-INF) [Fig. 3.14c]	Rewiring [11] (rewHier) [Fig. 3.14d]
$\mu F_1(\uparrow)$	77.04 (0.18)	78.00 (0.09)	79.42 (0.12)	**81.24 (0.08)**
$MF_1(\uparrow)$	77.94 (0.04)	78.20 (0.01)	79.82 (0.07)	**81.94 (0.04)**

The table shows mean and (standard deviation) in bracket across five runs. Best performing models are highlighted in bold

The results of classification performance is shown in Table 3.1. It can be seen that using these modified hierarchies substantially improves the classification performance in comparison to the baseline expert-defined hierarchy. On comparing the clustered, flattened, and rewiring hierarchies, the classification performance obtained from using the rewired hierarchy is found to be significantly better than the flattened and clustered hierarchy. This is because rewired hierarchy can resolve inconsistencies by grouping together the classes from different hierarchical branches that are effective. Hierarchy generated using clustering completely ignores the expert-defined hierarchy information, which contains valuable prior knowledge for classification [17], whereas flattening-based approaches cannot group together the classes from different hierarchical branches (e.g., *soc.religion.christian* and *religion.misc*).

3.4.2 Accuracy Comparisons: Flat Measures

Table 3.2 shows the μF_1 and MF_1 performance comparison of expert-defined hierarchy against clustered, flattened, and rewired hierarchy. Since, case study shows rewiring approaches as the best, experimental results are also included for rewiring approach proposed by Tang et al. [17] with minor modification. For reducing the number of operations (and hence hierarchy evaluations), hierarchy modifications are to the hierarchy branches where maximum classification errors are encountered. This modified approach is referred as T-Easy in the table.

The rewiring approaches consistently outperform other approaches for all the datasets across all metrics. For image datasets (CLEF, DIATOMS), the relative performance improvement is comparatively larger with performance improvement up to 11% using MF_1 scores in comparison to the baseline TD-LR method.

To further quantify the performance gain of rewiring approaches, the table shows the results of statistical significance test. Tests are performed between rewiring approaches and the next best performing approach, Global-INF.

In Table 3.2 results with p-values < 0.01 and < 0.05 are denoted by ▲ and △, respectively. Sign tests were used for μF_1 [20] and nonparametric Wilcoxon rank test for MF_1 comparing the F_1 scores obtained per class for the rewiring approaches against Global-INF. Both, the rewiring approaches significantly outperform the Global-INF method across the different datasets.

Table 3.2 μF_1 and MF_1 performance comparison using different hierarchy modification approaches

Dataset	Evaluation metrics	Baseline TD-LR	Clustering agglomerative [7]	Flattening Global-INF [10]	Rewiring methods T-Easy [17]	Rewiring methods rewHier [11]
CLEF	$\mu F_1(\uparrow)$	72.74	73.24	77.14	78.12	78.00
	$MF_1(\uparrow)$	35.92	38.27	46.54	**48.83▲**	47.10▲
DIATOMS	$\mu F_1(\uparrow)$	53.27	56.08	61.31	**62.34▲**	62.05▲
	$MF_1(\uparrow)$	44.46	44.78	51.85	**53.81▲**	52.14▲
IPC	$\mu F_1(\uparrow)$	49.32	49.83	52.30	53.94△	**54.28△**
	$MF_1(\uparrow)$	42.51	44.50	45.65	**46.10△**	46.04△
DMOZ-SMALL	$\mu F_1(\uparrow)$	45.10	45.94	46.61	NS	**48.25△**
	$MF_1(\uparrow)$	30.65	30.75	31.86	NS	**32.92▲**
DMOZ-2010	$\mu F_1(\uparrow)$	40.22	NS	42.37	NS	**43.10**
	$MF_1(\uparrow)$	28.37	NS	30.41	NS	**31.21**
DMOZ-2012	$\mu F_1(\uparrow)$	50.13	NS	50.64	NS	**51.82**
	$MF_1(\uparrow)$	29.89	NS	30.58	NS	**31.24**

▲ (△) indicates that improvements are statistically significant with 0.01 (0.05) significance level. Sign test and nonparametric Wilcoxon rank test were used for statistical evaluation of μF_1 and MF_1 scores, respectively. Tests are performed between rewiring approaches and the next best performing method, Global-INF. These statistical tests are not performed on DMOZ-2010 and DMOZ-2012 datasets because true labels are not available from the online evaluation system. "NS" denotes not scalable. Best performing models are highlighted in bold

On comparing two rewiring approaches, *rewHier* approach shows competitive classification performance in comparison to the T-Easy approach. For smaller datasets, the T-Easy approach has better performance because it searches for the optimal hierarchy in the hierarchical space. However, the main drawback of the T-Easy approach is that it requires computationally expensive learning-based evaluations for reaching the optimal hierarchy making it intractable for large, real-world classification benchmarks such as DMOZ (see detailed discussion in Runtime Comparisons).

3.4.3 Accuracy Comparisons: Hierarchical Measures

Hierarchical evaluation measures such as hF_1 computes errors for misclassified examples based on the definition of a defined hierarchy. Table 3.3 shows the hF_1 scores for the rewiring approaches and next best approach, Global-INF, evaluated over the original and the modified hierarchy. The rewiring approaches shows the best performance for all the datasets because it is able to restructure the hierarchy based on the dataset that is better suited for classification.

Table 3.3 hF_1 performance comparison using different hierarchy modification approaches

Dataset	Hierarchy used	Flattening Global-INF [10]	Rewiring methods T-Easy [17]	rewHier [11]
CLEF	Original	79.06	81.43	80.14
	Modified	80.87	81.82	81.28
DIATOMS	Original	62.80	64.28	63.24
	Modified	63.88	66.35	64.27
IPC	Original	64.73	67.23	68.34
	Modified	66.29	68.10	68.36
DMOZ-SMALL	Original	63.37	NS	66.18
	Modified	64.97	NS	66.30
DMOZ-2012	Original	73.19	NS	74.21

For DMOZ-2010 dataset hF_1 score is not available from the on-line evaluation system and for DMOZ-2012 dataset modified hierarchy is not supported

Moreover, new hierarchy has better performance over original hierarchy because restructuring leads to less misclassifications, resulting in hF_1 score improvement.

3.4.4 *Runtime Comparisons*

Table 3.4 shows the training times of the different approaches. From Table 3.4 it can be seen that TD-LR takes the least time as there is no overhead associated with modifying the hierarchy, followed by the Global-INF model which requires retraining of models after hierarchy flattening. Rewiring approaches are most expensive because of the compute-intensive task of either performing similarity computation in *rewHier* approach or multiple hierarchy evaluations using the T-Easy approach. The T-Easy method takes the longest time due to large number of expensive hierarchy evaluations after each elementary operations until the optimal hierarchy is reached. Table 3.5 shows the number of elementary operations executed using the T-Easy and the *rewHier* approach. It can be seen that T-Easy approach performs large number of operations even for smaller datasets (e.g., 412 operations for IPC datasets in comparison to 42 for the *rewHier*).

3.4.5 *Level-Wise Error Analysis*

Table 3.6 shows the level-wise error analysis for TD-LR approach with the original and rewired hierarchy (*rewHier*) on CLEF and DMOZ-SMALL datasets. It can be

Table 3.4 Total training time (in minutes)

Dataset	TD-LR	Flattening Global-INF [10]	Rewiring approaches T-Easy [17]	rewHier [11]
CLEF	2.5	3.5	59	7.5
DIATOMS	8.5	10	268	24
IPC	607	830	26,432	1284
DMOZ-SMALL	52	65	NS	168
DMOZ-2010	20,190	25,600	NS	42,000
DMOZ-2012	50,040	63,000	NS	94,800

Table 3.5 Number of elementary operation executed for rewiring approaches

# executed elementary operation for hierarchy modification	Dataset CLEF	DIATOMS	IPC
T-Easy [17] (promote, demote, merge)	52	156	412
rewHier [11] (node creation and deletion, parent-child rewiring)	25	34	42

Table 3.6 Level-wise error analysis for TD-LR approach on CLEF and DMOZ-SMALL with original and rewired hierarchy (*rewHier*)

Dataset	Level no.	Original Error (↓)	Original # ME	rewHier Error (↓)	rewHier # ME
CLEF	L-1	21.27 (0.63)	214	19.68 (0.18)	198
	L-2	07.71 (0.42)	240	05.85 (0.16)	212
	L-3	11.30 (0.16)	274	05.66 (0.31)	222
DMOZ-SMALL	L-1	42.47 (0.32)	789	38.97 (0.13)	724
	L-2	14.45 (0.62)	921	11.98 (0.18)	826
	L-3	15.14 (0.34)	972	14.72 (0.22)	919
	L-4	12.32 (0.02)	1001	09.26 (0.14)	934
	L-5	15.66 (0.05)	1020	12.13 (0.24)	961

The table shows mean and (standard deviation) of error rate across five runs. # ME denotes the average number of misclassified examples up to that level

seen from the table that rewired hierarchy makes fewer errors at each level which results in less error propagation and better overall classification performance. The table also shows the average number of misclassified examples (# ME) at each level in the table for better understanding the difference between the performance of these two hierarchies.

Table 3.7 Performance comparison with flat method

	Flat method		TD-LR				HierCost [3]			
	LR		Expert defined		rewHier		Expert defined		rewHier	
Dataset	MF_1	hF_1	MF_1	hF_1	MF_1	hF_1	MF_1	hF_1	MF_1	hF_1
CLEF	51.31	80.58	35.92	74.52	47.10	80.14	52.30	82.18	**54.20**	**84.42**
DIATOMS	54.17	63.50	44.46	56.15	52.14	63.24	54.16	64.13	**55.78**	**66.31**
IPC	45.74	64.00	42.51	62.57	46.04	62.57	50.10	68.45	**51.04**	**69.43**
DMOZ-SMALL	30.80	60.87	30.65	63.14	32.92	66.18	32.98	65.58	**33.43**	**66.30**
DMOZ-2010	27.06	53.94	28.37	54.82	29.48	56.43	29.81	58.24	**30.35**	**58.93**
DMOZ-2012	27.04	66.45	28.54	68.12	29.94	69.00	29.78	69.74	**30.27**	**70.21**

Best performing models are highlighted in bold

3.4.6 Comparison: Rewiring Against Flat and HierCost

Table 3.7 shows the MF_1 and hF_1 performance comparisons of flat LR, TD-LR, and HierCost [3] approach. The HC approaches are trained using the expert-defined hierarchy and compared to the one trained with the rewired hierarchy (*rewHier*).

From Table 3.7 it can be observed that the use of *rewHier* to train the TD-LR and HierCost improves the classification performance in comparison to using the expert-defined hierarchy. The HierCost approach in combination with *rewHier* outperforms all the methods on all the datasets. Moreover, TD-LR approach in combination with *rewHier* outperforms the flat approach for the large-scale DMOZ datasets with large numbers of rare categories. Rare categories benefit from utilization of the hierarchical relationships, and using the corrected hierarchy improves the accuracy of both the HC approaches.

Figure 3.15 presents the percentage of classes improved for TD-LR and HierCost HC approaches in comparison to the flat approach on DMOZ datasets containing rare categories. From the figure it can be observed that both the HC approaches significantly outperform the flat approach irrespective of the hierarchy being used. Specifically, >65% of the rare categories classes shows improved performance with the modified *rewHier* hierarchy. Moreover, the HierCost approach consistently outperforms the TD-LR approach because HierCost penalizes the misclassified instances based on the assignment within the hierarchy.

In terms of prediction runtime, the TD approaches outperform the flat and HierCost approaches. The flat and HierCost models invoke all the classifiers trained for the leaf nodes to make a prediction decision. For the DMOZ-2012 dataset, the flat and HierCost approaches take ~220 min for predicting the labels of test instances, whereas the TD-LR model is 3.5 times faster on the same hardware configuration.

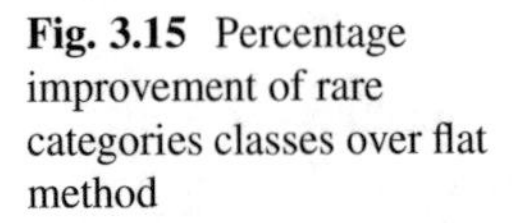

Fig. 3.15 Percentage improvement of rare categories classes over flat method

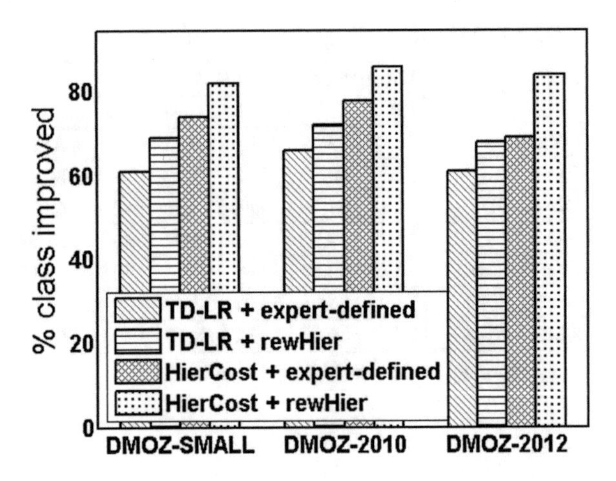

3.5 Summary of the Chapter

This chapter covers the various methodology for hierarchy modification that is useful for HC. In particular, flattening, rewiring, and clustering approaches have been discussed. These approaches are robust and can be adapted to work in conjunction with any state-of-the-art HC approaches in the literature that utilize hierarchical relationships.

References

1. Babbar, R., Partalas, I., Gaussier, E., Amini, M.R.: Maximum-margin framework for training data synchronization in large-scale hierarchical classification. In: Neural Information Processing, pp. 336–343 (2013)
2. Cai, L., Hofmann, T.: Hierarchical document categorization with support vector machines. In: Proceedings of the thirteenth ACM International Conference on Information and Knowledge Management, pp. 78–87 (2004)
3. Charuvaka, A., Rangwala, H.: Hiercost: Improving large scale hierarchical classification with cost sensitive learning. In: Joint European Conference on Machine Learning and Knowledge Discovery in Databases (ECML/PKDD), pp. 675–690 (2015)
4. Dumais, S., Chen, H.: Hierarchical classification of web content. In: Proceedings of the 23rd annual International ACM SIGIR Conference on Research and Development in Information Retrieval, pp. 256–263 (2000)
5. Gao, T., Koller, D.: Discriminative learning of relaxed hierarchy for large-scale visual recognition. In: Proceedings of the International Conference on Computer Vision (ICCV), pp. 2072–2079 (2011)
6. Koller, D., Sahami, M.: Hierarchically classifying documents using very few words. In: Proceedings of the Fourteenth International Conference on Machine Learning (ICML), pp. 170–178 (1997)
7. Li, T., Zhu, S., Ogihara, M.: Hierarchical document classification using automatically generated hierarchy. Journal of Intelligent Information Systems **29**(2), 211–230 (2007)

8. Liu, T.Y., Wan, H., Qin, T., Chen, Z., Ren, Y., Ma, W.Y.: Site abstraction for rare category classification in large-scale web directory. In: Special interest tracks and posters of the 14th International Conference on World Wide Web, pp. 1108–1109 (2005)
9. McCallum, A., Rosenfeld, R., Mitchell, T.M., Ng, A.Y.: Improving text classification by shrinkage in a hierarchy of classes. In: Proceedings of the 15th International Conference on Machine Learning (ICML), vol. 98, pp. 359–367 (1998)
10. Naik, A., Rangwala, H.: Inconsistent node flattening for improving top-down hierarchical classification. In: IEEE International Conference on Data Science and Advanced Analytics (DSAA), pp. 379–388 (2016)
11. Naik, A., Rangwala, H.: Improving large-scale hierarchical classification by rewiring: A data-driven filter based approach. Journal of Intelligent Information Systems (JIIS) pp. 1–24 (2018)
12. Nitta, K.: Improving taxonomies for large-scale hierarchical classifiers of web documents. In: Proceedings of the 19th ACM International Conference on Information and Knowledge Management, pp. 1649–1652 (2010)
13. Punera, K., Rajan, S., Ghosh, J.: Automatically learning document taxonomies for hierarchical classification. In: Special interest tracks and posters of the 14th International Conference on World Wide Web, pp. 1010–1011 (2005)
14. Qi, X., Davison, B.D.: Hierarchy evolution for improved classification. In: Proceedings of the 20th ACM International Conference on Information and Knowledge Management, pp. 2193–2196 (2011)
15. Steinbach, M., Ertöz, L., Kumar, V.: The challenges of clustering high dimensional data. In: New directions in statistical physics, pp. 273–309. Springer (2004)
16. Sun, A., Lim, E.P.: Hierarchical text classification and evaluation. In: Proceedings of the IEEE International Conference on Data Mining (ICDM), pp. 521–528 (2001)
17. Tang, L., Zhang, J., Liu, H.: Acclimatizing taxonomic semantics for hierarchical content classification. In: Proceedings of the 12th ACM SIGKDD International Conference on Knowledge Discovery and Data mining, pp. 384–393 (2006)
18. Wang, X.L., Lu, B.L.: Flatten hierarchies for large-scale hierarchical text categorization. In: Proceedings of the fifth International Conference on Digital Information Management (ICDIM), pp. 139–144 (2010)
19. Xiao, L., Zhou, D., Wu, M.: Hierarchical classification via orthogonal transfer. In: Proceedings of the 28th International Conference on Machine Learning (ICML), pp. 801–808 (2011)
20. Yang, Y., Liu, X.: A re-examination of text categorization methods. In: Proceedings of the 22nd annual International ACM SIGIR Conference on Research and Development in Information Retrieval, pp. 42–49 (1999)
21. Zamir, O., Etzioni, O.: Web document clustering: A feasibility demonstration. In: Proceedings of the 21st annual international ACM SIGIR conference on Research and development in information retrieval, pp. 46–54 (1998)
22. Zimek, A., Buchwald, F., Frank, E., Kramer, S.: A study of hierarchical and flat classification of proteins. IEEE/ACM Transactions on Computational Biology and Bioinformatics **7**(3), 563–571 (2010)

Chapter 4
Large-Scale Hierarchical Classification with Feature Selection

LSHC involves dataset consisting of thousands of classes and millions of training instances with high-dimensional features posing several big data challenges. Feature selection that aims to select the subset of discriminant features is an effective strategy to deal with large-scale problem. It speeds up the training process, reduces the prediction time, and minimizes the memory requirements by compressing the total size of learned model weight vectors. Majority of the studies have also shown feature selection to be competent and successful in improving the classification accuracy by removing irrelevant features. In this chapter, we investigate various filter-based feature selection methods for dimensionality reduction to solve the LSHC problem.

4.1 Introduction

Many LSHC approaches have been developed in the past to deal with the various "big data" challenges by (1) training faster models, (2) quickly predicting class labels, and (3) minimizing memory usage. For example, Gopal et al. [3] proposed the log-concavity bound that allows parallel training of model weight vectors across multiple computing units. This achieves significant speedup along with added flexibility of storing model weight vectors at different units. However, the memory requirement is still large (~26 GB for DMOZ-2010 dataset) which requires complex distributed hardware for storage and implementation. Alternatively, Map-Reduce based formulation of learning model is introduced [4, 8] which is scalable but have software/hardware dependencies that limits the applicability of this approach.

To minimize the memory requirements, one of the popular strategies is to incorporate the feature selection in conjunction with model training [5, 18]. The main intuition behind these approaches is to squeeze the high-dimensional features

A. Naik, H. Rangwala, *Large Scale Hierarchical Classification: State of the Art*, SpringerBriefs in Computer Science, https://doi.org/10.1007/978-3-030-01620-3_4

into lower dimensions. This allows the model to be trained on low-dimensional features only, significantly reducing the memory usage while retaining (or improving) the classification accuracy. This is possible because only subset of features are beneficial to discriminate between classes at each node in the hierarchy. For example, to distinguish between subclass *chemistry* and *physics* that belongs to class *science*, features like *chemicals*, *mixtures*, *velocity*, and *acceleration* are important, whereas features like *coach*, *memory*, and *processor* are irrelevant.

In this chapter, we evaluate different filter-based feature selection methods for solving LSHC problem. Feature selection serves as the preprocessing step in HC learning framework prior to training models. Any methods developed for solving HC problem can be integrated with the selected features, providing flexibility in choosing the HC algorithm of users' choice along with computational efficiency and storage benefits. Based on procedure followed for selecting relevant number of features at each node, two different formulations are discussed in this chapter: (1) global feature selection (Global FS) and (2) adaptive feature selection (Adaptive FS).

4.2 Feature Selection Overview

There have been several studies focused on feature selection methods for the flat classification problem [1, 6, 7, 10, 14, 17]. However, very few work emphasize on feature selection for HC problem that is limited to small number of categories [11, 15]. Figure 4.1 demonstrates the importance of feature selection for hierarchical settings where only the relevant features are chosen at each of the decision (internal) nodes. More details about the figure will be discussed in Sect. 4.3 (case study).

Feature selection aims to find a subset of highly discriminant features that minimize the error rate and improve the classifier performance. Based on the approach adapted for selecting features, two broad categories of feature selection exist, namely, wrapper and filter-based methods. Wrapper approaches evaluate the fitness of the selected features using the intended classifier. Although many different wrapper-based approaches have been proposed, these methods are not suitable for large-scale problems due to the expensive evaluation needed to select the subset of features [14]. On the contrary, filter approaches select the subset of features based on the certain measures or statistical properties that do not require the expensive evaluations. This makes the filter-based approach a natural choice for large-scale problem. Hence, in this book we have focused on various filter-based approaches for solving HC problem (discussed in Sect. 4.2.1). In literature, the third category referred as embedded approaches has also been proposed which is a hybrid of the wrapper and filter methods. However, these approaches have not been shown to be efficient for large-scale classification [14], and hence, we do not focus on hybrid methods in this book.

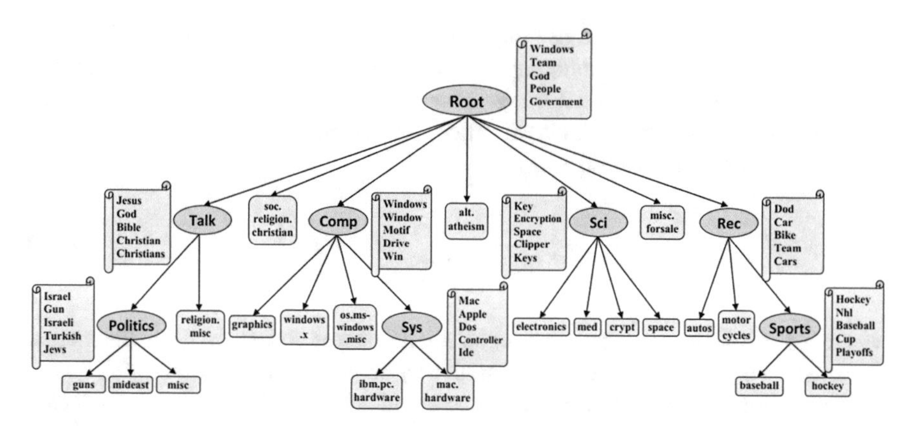

Fig. 4.1 Figure demonstrating the importance of feature selection for HC. Green color (sticky note) represents the top five best features selected using Gini index feature selection method at each internal node. Internal nodes are represented by orange color (elliptical shape), and leaf nodes are represented by blue color (rectangular shape)

4.2.1 Feature Selection Approaches

The focus of our study in this chapter is on filter-based feature selection methods which are scalable for large-scale datasets. In this section, we present four feature selection approaches that are used for evaluation purposes.

Gini Index It is one of the most widely used methods to compute the best split (ordered feature) in the decision tree induction algorithm [9]. Realizing its importance, it was extended for the multi-class classification problem [12]. In our case, it measure the feature's ability to distinguish between different leaf categories (classes). Gini index of i-th feature f_i with L classes can be computed as shown in Eq. (4.1).

$$\textbf{Gini-Index}(f_i) = 1 - \sum_{k=1}^{L} \left(p(k|f_i)\right)^2 \tag{4.1}$$

where $p(k|f_i)$ is the conditional probability of class k given feature f_i.

The smaller the value of Gini index, the more relevant and useful is the feature for classification. For HC problem, Gini index corresponding to all features is computed independently at each internal node, and the best subset of features ($S_{\mathscr{F}}$) are selected using a held-out validation dataset.

Minimal Redundancy Maximal Relevance (MRMR) This method incorporates the following two conditions for feature subset selection that are beneficial for classification:

1. Identify features that are mutually maximally dissimilar to capture better representation of entire dataset and
2. Select features to maximize the discrimination between different classes.

The first criterion referred as "minimal redundancy" selects features that carry distinct information by eliminating the redundant features. The main intuition behind this criterion is that selecting two similar features contains no new information that can assist in better classification. Redundancy information of feature set $\mathscr{F}$ can be computed using Eq. (4.2).

$$\mathfrak{R}_D = \left[\frac{1}{|S_{\mathscr{F}}|^2} \sum_{f_i, f_j \in S_{\mathscr{F}}} I(f_i, f_j) \right] \tag{4.2}$$

where $S_{\mathscr{F}} \subset \mathscr{F}$ denotes the subset of selected features and $I(f_i, f_j)$ is the mutual information that measure the level of similarity between features f_i and f_j [2].

The second criterion referred as "maximum relevance" enforces the selected features to have maximum discriminatory power for classification between different classes. Relevance of feature set $\mathscr{F}$ can be formulated using Eq. (4.3).

$$\mathfrak{R}_L = \left[\frac{1}{|S_{\mathscr{F}}|} \sum_{f_i \in S_{\mathscr{F}}} I(f_i, L) \right] \tag{4.3}$$

where $I(f_i, L)$ is the mutual information between the feature f_i and leaf categories L that captures how well the feature f_i can discriminate between different classes [10].

The combined optimization of Eqs. (4.2) and (4.3) leads to a feature set with maximum discriminatory power and minimum correlations among features. Depending on the strategy adapted for the optimization of these two objectives, different flavors exist. The first one referred as "mutual information difference (MRMR-D)" formulates the optimization problem as the difference between two objectives as shown in Eq. (4.4). The second one referred as "mutual information quotient (MRMR-Q)" formulates the problem as the ratio between two objectives and can be computed using Eq. (4.5).

$$\textit{MRMR-D} = \max_{S_{\mathscr{F}} \subseteq \mathscr{F}} (\mathfrak{R}_L - \mathfrak{R}_D) \tag{4.4}$$

$$\textit{MRMR-Q} = \max_{S_{\mathscr{F}} \subseteq \mathscr{F}} (\mathfrak{R}_L / \mathfrak{R}_D) \tag{4.5}$$

For HC problem, the best top $S_{\mathscr{F}}$ features (using a validation dataset) are selected for evaluating these methods.

Kruskal-Wallis (KW) This is a nonparametric statistical test that ranks the importance of each feature. As a first step, this method ranks all instances across all leaf categories L and computes the feature importance metric as shown in Eq. (4.6):

$$KW = (N-1) \frac{\sum_{i=1}^{L} n_i (\bar{r_i} - \bar{r})^2}{\sum_{i=1}^{L} \sum_{j=1}^{n_i} n_i (r_{ij} - \bar{r})^2} \tag{4.6}$$

Algorithm 1: Feature selection (FS)-based model learning for hierarchical classification (HC)

```
Data: Hierarchy H, input-output pairs (x(i), y(i))
Result: Learned model weight vectors:
        W = [W_1, W_2, ···, W_n], n ∈ 𝒩
1  W = φ;
2  /* 1st subroutine: feature selection */
3  for f_i ∈ ℱ do
4  |   Compute score (relevance) corresponding to feature f_i using feature selection algorithm
   |   mentioned in Sect. 4.2.1;
5  end
6  Select top k features based on score (and correlations) among features where best value of k
   is tuned using a validation dataset
7  /* 2nd subroutine: model learning using reduced feature set */
8  for n ∈ 𝒩 do
9  |   /* learn models for discriminating child at node n */
10 |   Train optimal multi-class classifiers W_n at node n using reduced feature set;
11 |   /* update model weight vectors */
12 |   W = [W, W_n];
13 end
14 return W
```

where n_i is the number of instances in i-th category, r_{ij} is the ranking of j-th instances in the i-th category, and $\bar{r}$ denotes the average rank across all instances.

It should be noted that using different feature results in different ranking and hence feature importance. The lower the value of computed score KW, the more relevant is the feature for classification.

4.2.2 Embedding Feature Selection into LSHC

Algorithm 1 presents method for embedding feature selection into the HC framework. It consists of two independent main subroutines: (1) a feature selection algorithm (discussed in Sect. 4.2.1) for deciding the appropriate set of features at each decision (internal) node and (2) a supervised learning algorithm for constructing a TD hierarchical classifier using reduced feature set. Feature selection serves as the preprocessing step in this framework which provides flexibility in choosing any HC algorithm.

There are two different approaches for choosing relevant number of features at each internal node $n \in \mathcal{N}$. The first approach referred as "global feature selection (Global FS)" selects the same number of features for all internal nodes in the hierarchy where the number of features are determined based on the entire validation dataset performance. The second approach, referred as "adaptive feature selection (Adaptive FS)," selects different number of features at each internal node to maximize the performance at that node. It should be noted that adaptive method only uses the validation dataset that exclusively belongs to the internal node n (i.e., descendant categories of node n). Computationally, both approaches are almost identical because model tuning and optimization require similar runtime which accounts for the major fraction of computation.

4.3 Experimental Results and Analysis

4.3.1 Case Study

To understand the quality of features selected at different internal nodes in the hierarchy, we perform case study on NG dataset. We choose this dataset because we have full access to feature information. Figure 4.1 demonstrates the results of top five features that are selected using best feature selection method, i.e., Gini index (refer to Figs. 4.2 and 4.3). We can see from the figure that selected features correspond to the distinctive attributes which help in better discrimination at particular node. For example, the features like *Dod (day of defeat or Department of Defense)*, *car*, *bike*, and *team* are important at node *Rec* to distinguish between the subclass *autos*, *motorcycles*, and *sports*, whereas other features like *windows*, *God*, and *encryption* are irrelevant. This analysis illustrates the importance of feature selection for top-down HC problem.

One important observation that we made in our study is that some of the features like *windows*, *God*, and *team* are useful for discrimination at multiple nodes in the hierarchy (associated with parent-child relationships). This observation conflicts with the assumption made in the work by Xiao et al. [18], which attempts to optimize the objective function by necessitating the child node features to be different from the features selected at the parent node.

4.3.2 Accuracy Comparison

Global FS Figures 4.2 and 4.3 shows the μF_1 and MF_1 comparison of LR models with l_1-*norm* and l_2-*norm* regularization combined with various feature selection methods discussed in Sect. 4.2.1, respectively. It can be seen that all feature selection methods (except Kruskal-Wallis) show competitive performance results in comparison to the full set of features for all the datasets. Overall, Gini index feature selection method has slightly better performance over other methods. MRMR methods have a tendency to remove some of the important features as redundant based on the minimization objective obtained from data-sparse leaf categories which may not be optimal and negatively influence the performance. The Kruskal-Wallis method shows poor performance because of the statistical properties that are obtained from data-sparse nodes [13].

On comparing the l_1-*norm* and l_2-*norm* regularized models of best feature selection method (Gini index) with all features, it can be seen that l_1-*norm* models

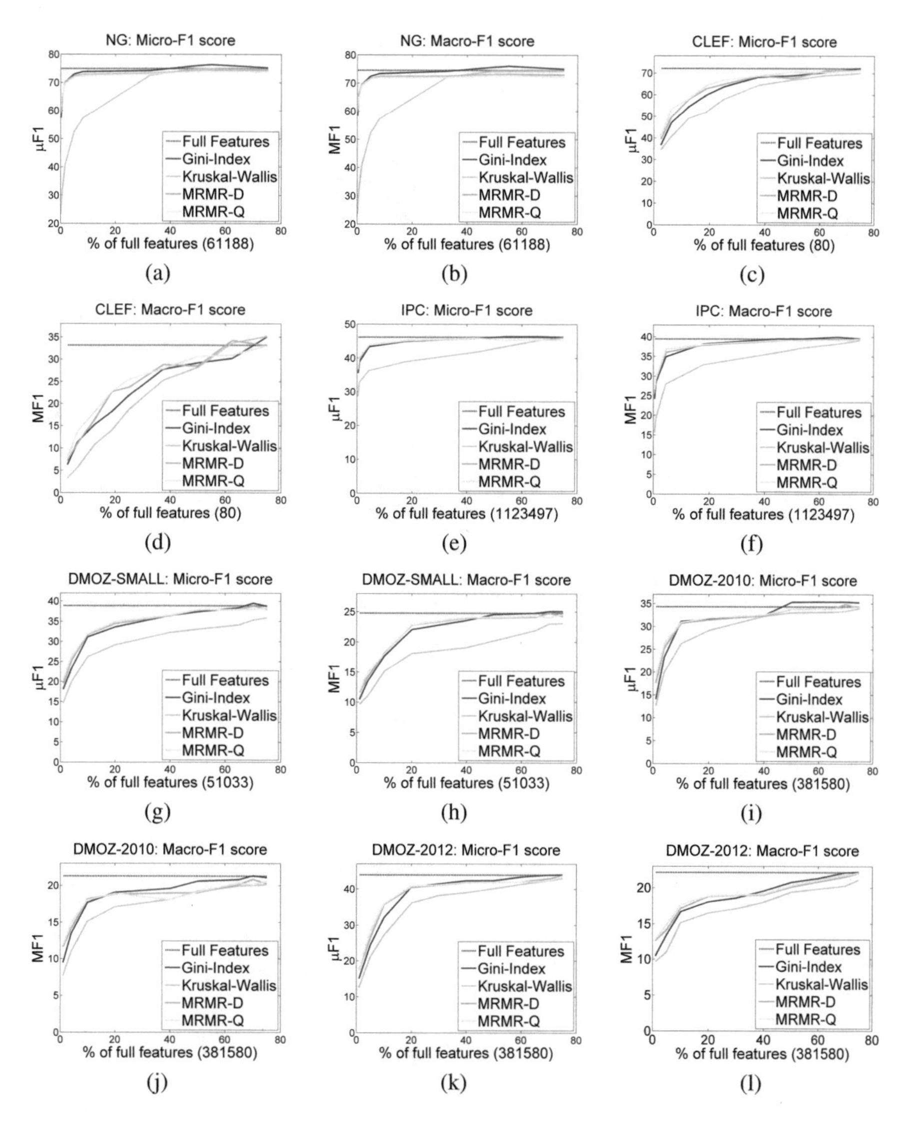

Fig. 4.2 Performance comparison of LR + l_1-*norm* models with varying percentage (%) of features selected using different feature selection (Global) methods. **(a)**–**(l)** shows the Micro-F1 and Macro-F1 performance on different datasets

have more performance improvement (especially for MF_1 scores) for all datasets, whereas for l_2-*norm* models, performance is almost similar without any significant loss. This is because l_1-*norm* assigns higher weight to the important predictor variables which results in more performance gain.

Since feature selection based on Gini index gives the best performance in the rest of the experiments, Gini index is used as the baseline for comparison purpose. Also, results for l_1-*norm* models only are shown due to space constraints.

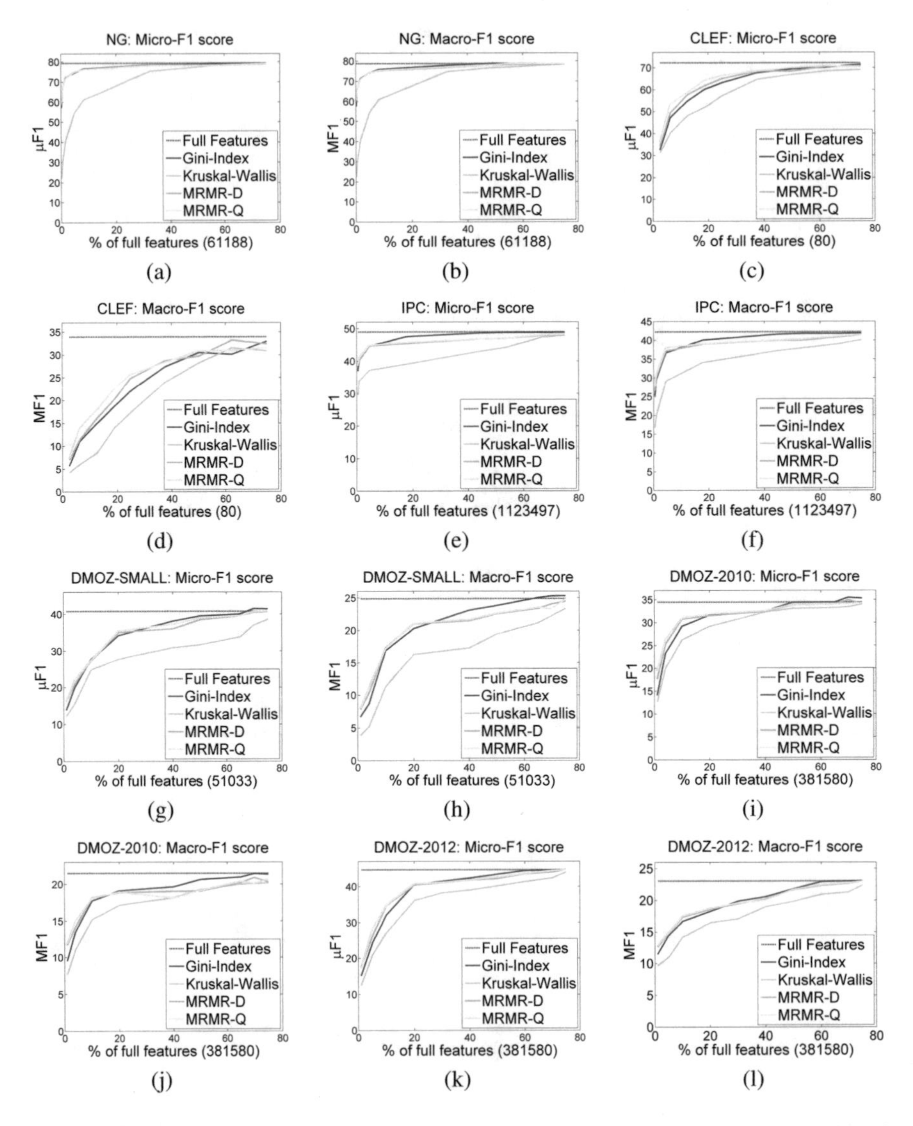

Fig. 4.3 Performance comparison of LR + l_2-norm models with varying percentage (%) of features selected using different feature selection (Global) methods. (**a**)–(**l**) shows the Micro-F1 and Macro-F1 performance on different datasets

Adaptive FS Table 4.1 shows the LR + l_1-*norm* model performance comparison of adaptive and global approaches for feature selection with all features. It can be seen from the table that adaptive approach-based feature selection gives the best performance for all the datasets (except μF_1 score of newsgroup dataset which has very few categories). For evaluating the performance improvement of models, statistical significance test is performed. Specifically, sign test is used for μF_1 [16]

Table 4.1 Performance comparison of adaptive and global approach for feature selection based on Gini index with all features

Dataset	Metric	Adaptive FS	Global FS	All features
NG	μF_1	76.16△	**76.39**△	74.94
	MF_1	**76.10**△	76.07△	74.56
CLEF	μF_1	**72.66**	72.27	72.17
	MF_1	**36.73**▲	35.07△	33.14
IPC	μF_1	**48.23**▲	46.35	46.14
	MF_1	**41.54**▲	39.52	39.43
DMOZ-SMALL	μF_1	**40.32**△	39.52	38.86
	MF_1	**26.12**▲	25.07	24.77
DMOZ-2010	μF_1	**35.94**	35.40	34.32
	MF_1	**23.01**	21.32	21.26
DMOZ-2012	μF_1	**44.12**	43.94	43.92
	MF_1	**23.65**	22.18	22.13

LR + l_1-$norm$ model is used for evaluation. Best performing methods are highlighted in bold
▲ (and △) indicates that improvements are statistically significant with 0.05 (and 0.1) significance level

Table 4.2 Comparison of memory requirements for LR + l_1-$norm$ model

	Adaptive FS		Global FS		All features	
Dataset	#parameters	Size	#parameters	Size	#parameters	Size
NG	982,805	4.97 MB	**908,820**	**3.64 MB**	1,652,076	6.61 MB
CLEF	**4715**	**18.86 KB**	5220	20.89 KB	6960	27.84 KB
IPC	**306,628,256**	**1.23 GB**	331,200,000	1.32 GB	620,170,344	2.48 GB
DMOZ-SMALL	**74,582,625**	**0.30 GB**	85,270,801	0.34 GB	121,815,771	0.49 GB
DMOZ-2010	**4,035,382,592**	**16.14 GB**	4,271,272,967	17.08 GB	6,571,189,180	26.28 GB
DMOZ-2012	**3,453,646,353**	**13.81 GB**	3,649,820,382	14.60 GB	4,866,427,176	19.47 GB

Best performing methods are highlighted in bold

and nonparametric Wilcoxon rank test for MF_1. Result with 0.05 (0.1) significance level is denoted by ▲ (△). Tests are between models obtained using feature selection methods and all set of features. Tests are not performed on DMOZ-2010 and DMOZ-2012 datasets because true predictions and class-wise performance score are not available from online web portal.

Statistical evaluation shows that although global approach is slightly better in comparison to full set of features, they are not statistically significant. On contrary, adaptive approach is much better with an improvement of ~2% in μF_1 and MF_1 scores which are statistically significant.

4.3.3 Memory Requirements

Table 4.2 shows the information about memory requirements for various models with full set of features and best set of features that are selected using global and adaptive feature selection. Up to 45% reduction in memory size is observed for all datasets to store the learned models. This is a huge margin in terms of memory requirements considering the models for large-scale datasets (such as DMOZ-2010 and DMOZ-2012) are difficult to fit in memory.

It should be noted that optimal set of features is different for global and adaptive methods for feature selection; hence they have different memory requirements. Overall, Adaptive FS is slightly better because it selects small set of features that are relevant for distinguishing data-sparse nodes present in CLEF, IPC, and the DMOZ datasets. Also, we would like to point out that Table 4.2 represents the memory required to store the learned model parameters only. In practice, 2–4 times more memory are required for temporarily storing the gradient values of model parameters that are obtained during the optimization process.

4.3.4 Runtime Comparison

Preprocessing Time Table 4.3 shows the preprocessing time needed to compute the feature importance using the different feature selection methods. The Gini index method takes the least amount of time since it does not require the interactions between different features to rank the features. The MRMR methods are computationally expensive due to the large number of pairwise comparisons between all the features to identify the redundancy information. On other hand, the Kruskal-Wallis method has overhead associated with determining ranking of each features with different classes.

Model Training Table 4.4 shows the total training time needed for learning models. As expected, feature selection requires less training time due to the less number of features that needs to be considered during learning. For smaller datasets such as NG and CLEF, improvement is not noticeable. However, for larger datasets with high dimensionality such as IPC, DMOZ-2010, and DMOZ-2012, improvement is much higher (up to $3\times$ order speedup). For example, DMOZ-2010 dataset training time reduces from 6524 min to mere 2258 min.

Prediction Time For the dataset with largest number of test instances, DMOZ-2012 takes 37 min to make predictions with feature selection as opposed to 48.24 min with all features using the top-down HC approach.

Figure 4.4 shows the training and prediction time comparison of large datasets (DMOZ-2010 and DMOZ-2012) between flat LR and the top-down HC approach with (and without) feature selection. The flat method is comparatively more expensive than the TD approach ($\sim$6.5 times for training and $\sim$5 times for prediction).

Table 4.3 Feature selection preprocessing time (in minutes)

	Feature selection method			
Dataset	Gini index	MRMR-D	MRMR-Q	Kruskal-Wallis
NG	**2.10**	5.33	5.35	5.42
CLEF	**0.02**	0.46	0.54	0.70
IPC	**15.2**	27.42	27.00	23.24
DMOZ-SMALL	**23.65**	45.24	45.42	34.65
DMOZ-2010	**614**	1524	1535	1314
DMOZ-2012	**818**	1824	1848	1268

Best performing methods are highlighted in bold

Table 4.4 Total training time (in minutes)

Dataset	Model	Feature selection (Gini index)	All features
NG	LR + l_1	0.75	0.94
	LR + l_2	0.44	0.69
CLEF	LR + l_1	0.50	0.74
	LR + l_2	0.10	0.28
IPC	LR + l_1	24.38	74.10
	LR + l_2	20.92	68.58
DMOZ-SMALL	LR + l_1	3.25	4.60
	LR + l_2	2.46	3.17
DMOZ-2010	LR + l_1	2258	6524
	LR + l_2	2132	6418
DMOZ-2012	LR + l_1	8024	19,374
	LR + l_2	7908	19,193

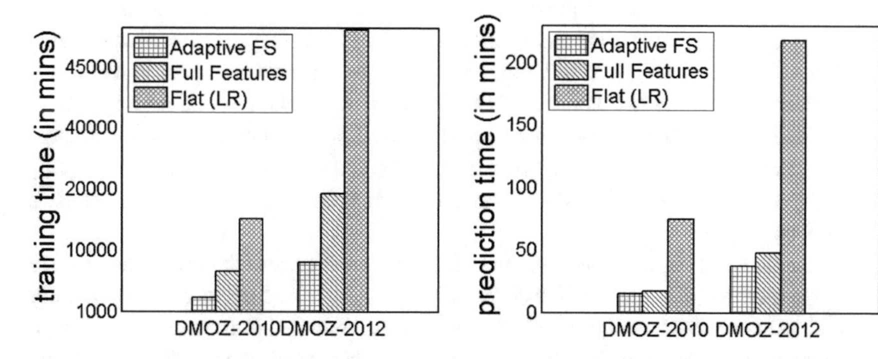

Fig. 4.4 Training and prediction runtime comparison of LR + l_1-*norm* model (in minutes)

Table 4.5 Performance comparison of LR + l_1-*norm* model with varying training size (number of instances) per class on newsgroup dataset

Dataset distribution	Train size (per class)	Feature selection (Gini index) μF_1	MF_1	All features μF_1	MF_1
Low distribution	5	**27.44** ▲ (0.4723)	**26.45** ▲ (0.4415)	25.74 (0.5811)	24.33 (0.6868)
	10	**37.69** △ (0.2124)	**37.51** ▲ (0.2772)	36.59 (0.5661)	35.86 (0.3471)
	15	**43.14** △ (0.3274)	**43.80** △ (0.3301)	42.49 (0.1517)	42.99 (0.7196)
	25	**52.12** ▲ (0.3962)	**52.04** ▲ (0.3011)	50.33 (0.4486)	50.56 (0.5766)
High distribution	50	**59.55** (0.4649)	59.46 (0.1953)	59.52 (0.3391)	**59.59** (0.1641)
	100	66.53 (0.0346)	66.42 (0.0566)	**66.69** (0.7321)	**66.60** (0.8412)
	200	70.60 (0.6068)	70.53 (0.5164)	**70.83** (0.7123)	**70.70** (0.6330)
	250	72.37 (0.4285)	72.24 (0.4293)	**73.06** △ (0.4732)	**72.86** (0.4898)

The table shows mean and standard deviation in bracket across five runs. ▲ (and △) indicates that improvements are statistically significant with 0.05 (and 0.1) significance level. Best performing methods are highlighted in bold

4.3.5 *Effect of Varying Training Size*

Table 4.5 shows the classification performance on newsgroup dataset with varying training dataset distribution. Models are tested by varying the training size (instances) per class (t_c) between 5 and 250. Each experiment is repeated five times by randomly choosing t_c instances per class. Moreover, adaptive method with Gini index feature selection is used for experiments. For evaluating the performance improvement of models, we perform statistical significance test (sign test for μF_1 and Wilcoxon rank test for MF_1). Results with 0.05 (0.1) significance level is denoted by ▲ (△).

It can be seen from Table 4.5 that for low distribution datasets, the feature selection method performs well and shows improvements of up to 2% (statistically significant) over the baseline method. The reason behind this improvement is that with low data distribution, feature selection methods prevent the models from over-fitting by selectively choosing the important features that helps in discriminating between the models of various classes. For datasets with high distribution, no significant performance gain is observed due to sufficient number of available training instances for learning models which prevent over-fitting when using all the features.

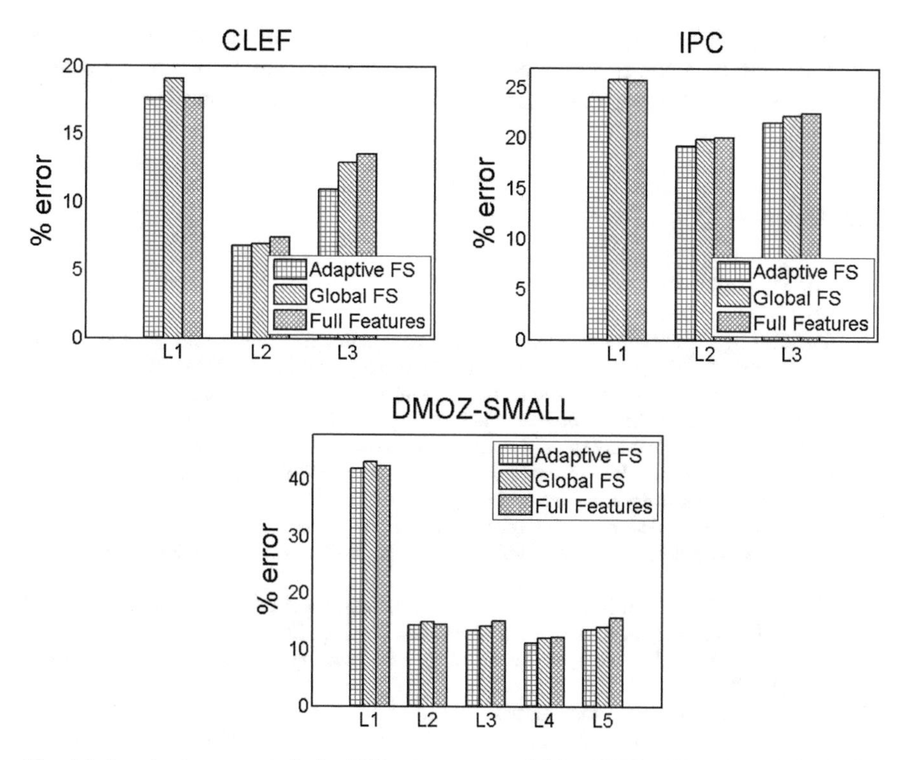

Fig. 4.5 Level-wise error analysis of LR + l_1-*norm* model for CLEF, IPC, and DMOZ-SMALL datasets

4.3.6 Level-Wise Error Analysis

Figure 4.5 shows the level-wise error analysis for CLEF, IPC, and DMOZ-SMALL datasets with or without feature selection. It can be seen that at topmost level more error is committed compared to the lower level. This is because at higher levels each of the children nodes that needs to be discriminated is the combination of multiple leaf categories which cannot be modeled accurately using the linear classifiers. Another observation is that adaptive feature selection gives best results at all levels for all datasets which demonstrates its ability to extract relevant number of features at each internal node (that belongs to different levels) in the hierarchy.

4.4 Summary of the Chapter

In this chapter, two different approaches for embedding feature selection into HC framework are discussed. For evaluation, four different easily parallelizable feature selection methods are discussed. Experimental evaluation shows that with feature selection significant improvement is achieved in terms of runtime performance (training and prediction) and memory requirements (especially for large-scale datasets) without affecting the accuracy of learned classification models.

References

1. Dash, M., Liu, H.: Feature selection for classification. Intelligent Data Analysis **1**(3), 131–156 (1997)
2. Ding, C., Peng, H.: Minimum redundancy feature selection from microarray gene expression data. Journal of bioinformatics and computational biology **3**(02), 185–205 (2005)
3. Gopal, S., Yang, Y.: Distributed training of large-scale logistic models. In: Proceedings of the 30th International Conference on Machine Learning (ICML), pp. 289–297 (2013)
4. Gopal, S., Yang, Y.: Recursive regularization for large-scale classification with hierarchical and graphical dependencies. In: Proceedings of the 19th ACM SIGKDD International Conference on Knowledge Discovery and Data mining, pp. 257–265 (2013)
5. Heisele, B., Serre, T., Prentice, S., Poggio, T.: Hierarchical classification and feature reduction for fast face detection with support vector machines. Pattern Recognition **36**(9), 2007–2017 (2003)
6. Joachims, T.: Text categorization with support vector machines: Learning with many relevant features. In: European Conference on Machine Learning, pp. 137–142 (1998)
7. Kohavi, R., John, G.H.: Wrappers for feature subset selection. Artificial intelligence **97**(1), 273–324 (1997)
8. Naik, A., Rangwala, H.: A ranking-based approach for hierarchical classification. In: IEEE International Conference on Data Science and Advanced Analytics (DSAA), pp. 1–10 (2015)
9. Ogura, H., Amano, H., Kondo, M.: Feature selection with a measure of deviations from poisson in text categorization. Expert Systems with Applications **36**(3), 6826–6832 (2009)
10. Peng, H., Long, F., Ding, C.: Feature selection based on mutual information criteria of max-dependency, max-relevance, and min-redundancy. IEEE Transactions on Pattern Analysis and Machine Intelligence **27**(8), 1226–1238 (2005)
11. Ristoski, P., Paulheim, H.: Feature selection in hierarchical feature spaces. In: International Conference on Discovery Science, pp. 288–300. Springer (2014)
12. Shang, W., Huang, H., Zhu, H., Lin, Y., Qu, Y., Wang, Z.: A novel feature selection algorithm for text categorization. Expert Systems with Applications **33**(1), 1–5 (2007)
13. Strobl, C., Zeileis, A.: Danger: High power! exploring the statistical properties of a test for random forest variable importance. In: In Proceedings of the 18th International Conference on Computational Statistics (2008)
14. Tang, J., Alelyani, S., Liu, H.: Feature selection for classification: A review. Data Classification: Algorithms and Applications p. 37 (2014)
15. Wibowo, W., Williams, H.E.: Simple and accurate feature selection for hierarchical categorisation. In: Proceedings of the 2002 ACM symposium on Document Engineering, pp. 111–118 (2002)
16. Yang, Y., Liu, X.: A re-examination of text categorization methods. In: Proceedings of the 22nd annual International ACM SIGIR Conference on Research and Development in Information Retrieval, pp. 42–49 (1999)
17. Zheng, Z., Wu, X., Srihari, R.: Feature selection for text categorization on imbalanced data. ACM Sigkdd Explorations Newsletter **6**(1), 80–89 (2004)
18. Zhou, D., Xiao, L., Wu, M.: Hierarchical classification via orthogonal transfer. In: Proceedings of the 28th International Conference on Machine Learning (ICML), pp. 801–808 (2011)

Chapter 5
Multi-task Learning

5.1 Introduction

Traditional supervised machine learning methods involve learning mapping function that can accurately map input data to output label. However, real-world datasets are complex, and we often encounter situations where multiple tasks (classes) are related to each other. For example, consider an email spam classification where the goal is to learn classification model to categorize email as spam or legitimate for each user. Spam email for one user might be related to the spam email for other users. Intuitively, it would seem that learning these related tasks jointly would help us uncover common knowledge and improve generalization performance. In fact, this intuition is supported by empirical evidence provided by recent developments in transfer learning (TL) [26] and multi-task learning (MTL) [7, 29].

MTL involves learning of multiple-related tasks, jointly. It seeks to improve the performance of each task by leveraging the relationships among the different tasks. It is an advanced concept of Single-Task Learning (STL), most widely used in classification. In STL, each task is considered to be independent, and the model parameters are learned independently, whereas in MTL, multiple tasks are learned simultaneously by utilizing task relatedness as shown in Fig. 5.1. The main intuition behind MTL is that the training signal present in related tasks can help each of the task learn better model parameters [3, 7]. It has been shown that MTL improves the performance of the model, especially when the number of training examples is less and when the tasks are related [4, 29].

> Sharing information between unrelated tasks in MTL might hurt the performance. This is due to the phenomenon known as negative transfer.

A. Naik, H. Rangwala, *Large Scale Hierarchical Classification: State of the Art*,
SpringerBriefs in Computer Science, https://doi.org/10.1007/978-3-030-01620-3_5

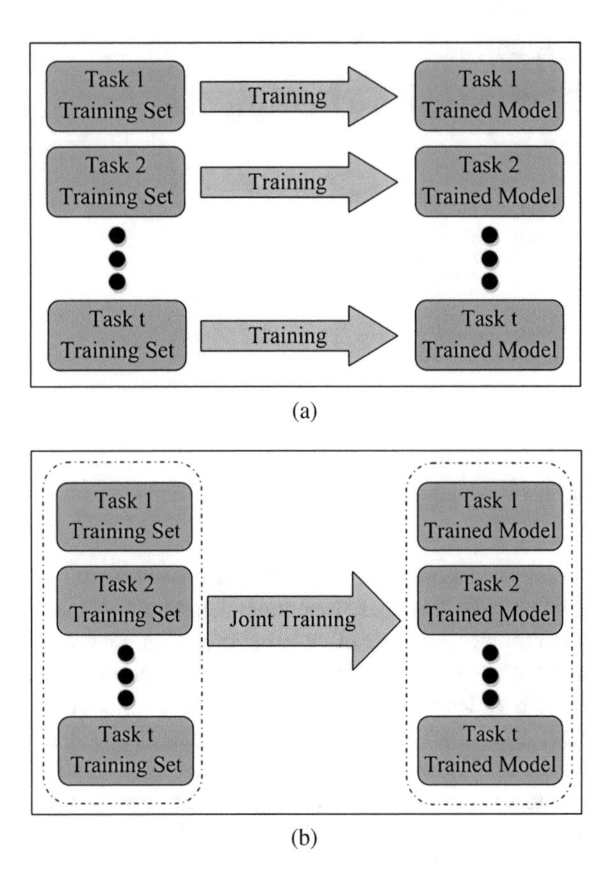

Fig. 5.1 Figure shows difference between (**a**) Single Task Learning (STL) and (**b**) Multi-Task Learning (MTL) model training. In STL, each task is learned independently whereas in MTL multiple tasks are learned jointly

MTL captures the intrinsic relatedness between the tasks and hence achieve better generalization performance, especially when the number of training examples is less. MTL has been shown to be effective and successfully applied across multiple applications including medical informatics [38], structural classification [9], sequence analysis [33], language processing [10], drug discovery [28], web image and video search [32], and many more [11, 14, 15, 34, 38]. There has been huge research exploration in the MTL research, and the literature review about the MTL can be found in Zhou et al. [37].

5.2 Multi-task Learning Problem Formulation

Given a training dataset with N input-output pairs $\{(\mathbf{x}(1), y(1)), (\mathbf{x}(2), y(2)), \cdots, (\mathbf{x}(N), y(N))\}$, the goal is to learn a mapping function $f : X \rightarrow Y$ between the input domain $\mathbf{x}(i) \in X$ and output domain $y(i) \in Y$. X and Y are input and output domains, respectively. The objective is to learn a model that minimizes the loss function on the training data while constraining the model complexity with a regularization penalty. The learning objective for each of the task t in the regularized

STL can be given as,

$$\min_{\mathbf{w}_t} \sum_{i=1}^{N} \underbrace{\mathscr{L}(\mathbf{w}_t, \mathbf{x}(i), y(i))}_{Empirical\ Loss} + \lambda \underbrace{\Omega(\mathbf{w}_t)}_{Regularization} \tag{5.1}$$

where $\mathbf{w}_t$ represents the model parameters for $t\text{-}th$ task. The regularization term controls the model complexity, thus safeguarding against the model over-fitting. Extension of STL is MTL which learns the model parameters of the related task together. In MTL we are given T tasks with training set defined for each of the $t = 1 \ldots T$ tasks, given by $(\mathbf{x}_t(i), y_t(i)) : i = 1, 2, \ldots, N_t$, and the combined learning objective is given by,

$$\min_{\mathbf{W}} \sum_{t=1}^{T} \sum_{i=1}^{N_t} \underbrace{\mathscr{L}\big(\mathbf{w}_t, \mathbf{x}_t(i), y_t(i)\big)}_{Empirical\ Loss} + \lambda \underbrace{\Omega(\mathbf{W})}_{Regularization} \tag{5.2}$$

where N_t is the number of training instances for the $t\text{-}th$ task, $\mathbf{w}_t$ denotes the model parameters for the $t\text{-}th$ task, $(\mathbf{x}_t(i), y_t(i))$ represents the $i\text{-}th$ input and output pair for $t\text{-}th$ task, and $\mathbf{W} = \{\mathbf{w}_t\}_{t=1}^{T}$ is the combined set of model parameters for all the related tasks. Various multi-task learning methods take this general approach to build combined models for many related tasks. In Evgeniou et al. [13] the model for each task is constrained to be close to the average of all the tasks. In multi-task feature learning and feature selection methods [2, 18, 21, 23], sparse learning based on lasso [31] is performed to select or learn a common set of features across many related tasks. However, a common assumption made by many methods [1, 13, 17] is that all tasks are equally related. This assumption does not hold in all situations.

Therefore, it is sensible to take the task relationships into account in MTL. Kato et al. [19] and Evgeniou et al. [12] propose formulations which use an externally provided task network or graph structure. However, these relationships might not be available and may need to be determined from the data. Clustered MTL approaches assume that tasks exhibit a group structure, which is not known a priori and seeks to learn the clusters of tasks that are learned together [16, 30, 36]. Another set of approaches, mostly based on Gaussian process models, learn the task covariance structure [6, 35] and are able to take advantage of both positive and negative correlations between the tasks.

In this chapter, we have focused on the MTL-based models for the purpose of multi-class classification using multiple hierarchies.

5.3 Multi-task Learning Using Multiple Hierarchies

Hierarchies have become popular structures for data representation. It is so common that sometimes multiple hierarchies classify similar data. For example, in protein structure classification, several hierarchies exist for organizing proteins

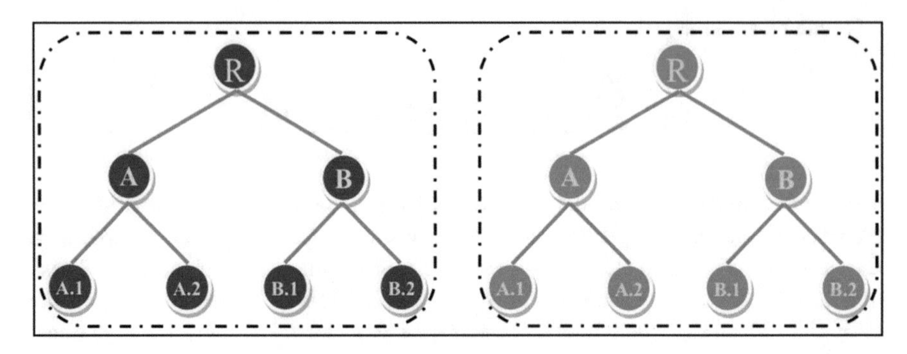

Fig. 5.2 Single Hierarchy Multi-Task Learning: related tasks are identified individually within each hierarchy

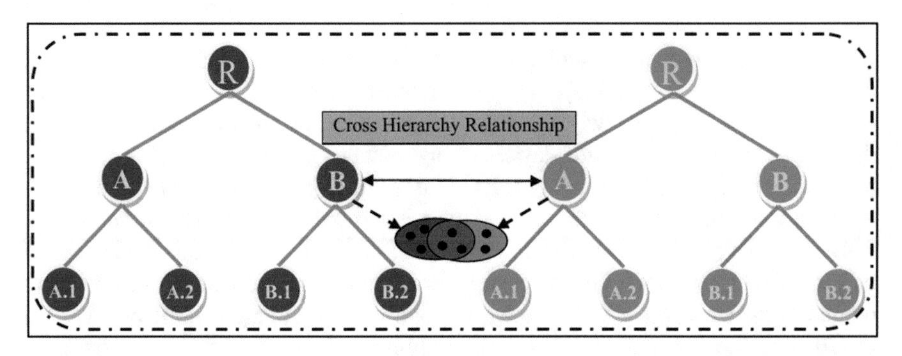

Fig. 5.3 Multiple Hierarchy Multi-Task Learning: related tasks are identified across multiple hierarchies for joint learning

based on curation process or 3D structure. Another example is web-page classification where several hierarchies (such as DMOZ and Wikipedia) exist for categorizing web-pages. In order to improve classification performance, we need to identify related tasks and learn models together. Based on how we leverage multiple hierarchies for finding related tasks, there are two different ways of MTL [8].

1. Single Hierarchy Multi-Task Learning (SHMTL)—In SHMTL, each hierarchy is considered independently for MTL (Fig. 5.2). Relationship between tasks within a hierarchy is combined individually.
2. Multiple Hierarchy Multi-Task Learning (MHMTL)—In MHMTL, multiple hierarchies are considered together for MTL (Fig. 5.3). Relationship between tasks from different hierarchies is extracted using common instances or similarity metric such as *k*NN.

5.4 Performance Comparison of STL, SHMTL, and MHMTL

To illustrate the effectiveness of MTL methods, in this section we will provide and discuss some of the experimental results from the paper [8]. Experiments have been performed on protein structures hierarchical database—structural classification of proteins (SCOP) [22] and class, architecture, topology, and homologous (CATH) [24] superfamily. Both these databases classify proteins into four major levels. For experimental evaluations, the following MTL methods have been used:

1. Sparse MTL—In this method, it is assumed that across all the tasks, only a subset of the features are important for classification [1, 21, 23]. The objective function for sparse MTL can be represented as,

$$\min_{\mathbf{W}} \sum_{t=1}^{T} \sum_{i=1}^{N_t} \mathscr{L}(\mathbf{w}_t, \mathbf{x}_t(i), y_t(i)) + \lambda ||\mathbf{W}||_{2,1} \tag{5.3}$$

where $||\mathbf{W}||_{2,1}$ denotes the $l_{2,1}$-*norm*.

$l_{2,1}$-*norm* is defined as the l_1-*norm* of the vector of l_2-*norm* over the weights associated with a particular input dimension.

2. Graph MTL—In this method, edge relationships in the hierarchy are exploited in the regularization. This method minimizes the difference between all pairs of related tasks (connected by edges) unlike [13] where all tasks model parameters are forced to be similar to average of all tasks.

$$\min_{\mathbf{W}} \sum_{t=1}^{T} \sum_{i=1}^{N_t} \mathscr{L}(\mathbf{w}_t, \mathbf{x}_t(i), y_t(i)) + \lambda \sum_{(p,q)\in\mathscr{E}} ||\mathbf{w}_p - \mathbf{w}_q||_2^2 \tag{5.4}$$

where $(p, q) \in \mathscr{E}$ denotes the edge in the hierarchy.

3. Trace MTL—In this method, trace norm is used in the regularization as shown in Eq. (5.5). Trace norm forces $\mathbf{W}$ to share a low-dimensional subspace, therefore inducing correlations between the tasks. This formulation has been extensively studied in the paper by Tong et al. [27].

$$\min_{\mathbf{W}} \sum_{t=1}^{T} \sum_{i=1}^{N_t} \mathscr{L}(\mathbf{w}_t, \mathbf{x}_t(i), y_t(i)) + \lambda ||\mathbf{W}||_* \tag{5.5}$$

where $||\mathbf{W}||_*$ denotes the trace norm.

4. Graph + Trace MTL—This method is a combination of Graph MTL and Trace MTL as shown in Eq. (5.6).

$$\min_{\mathbf{W}} \sum_{t=1}^{T} \sum_{i=1}^{N_t} \mathscr{L}\big(\mathbf{w}_t, \mathbf{x}_t(i), y_t(i)\big) + \lambda_1 \sum_{(p,q)\in\mathscr{E}} ||\mathbf{w}_p - \mathbf{w}_q||_2^2 + \lambda_2 ||\mathbf{W}||_* \quad (5.6)$$

In all MTL methods discussed above, logistic loss was used as the loss function, and an *l2-norm* is added to the regularization objective. *l2-norm* helps in controlling the magnitude of model parameters.

5.4.1 Experimental Analysis

Data Representation Spectrum kernel features method [20] (which uses contiguous subsequences of some fixed length k, also known as k-mers) is used to represent protein sequences into feature vectors of fixed length. Experiments were performed by setting $k = 3$ which resulted in 8000 features.

Extracting Cross-hierarchy Relationships Edges are derived between two hierarchies using consistent protein domains. If a domain is classified by two nodes, then an edge is added between them.

Results For comparison of different methods, only level 3 (L3) and level 4 (L4) classification results are included in the results. Figure 5.4 shows the performance improvements of different MTL methods over STL method. In general MTL methods have better performance than STL with only exception in case of sparse MTL methods. This is because sparse formulation attempts to extract sparse features across all the tasks. It assumes a uniform relationship between all the tasks.

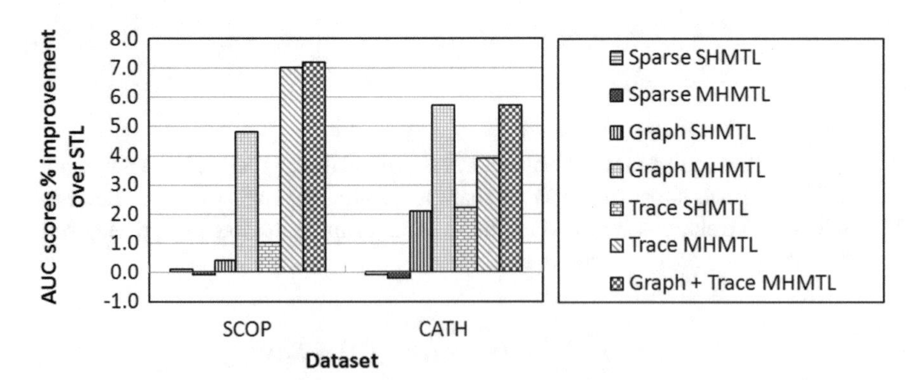

Fig. 5.4 Comparison of STL against different MTL methods. Figure shows AUC scores % improvements of MTL methods over STL using L3 along with auxiliary L4 tasks in training

However, it is unlikely that any subset for the features derived from the protein sequences are more important for the classification. Similar performance is reported with considering L3 nodes only.

5.5 Performance Comparison of STL, MTL, TL, and SSL

In this section, we provide experimental results for STL, MTL, transfer learning (TL), and semi-supervised learning (SSL). For better understanding the material, we first discuss the problem that we are trying to solve followed by different formulations that have been used for comparison. Finally, we provide some of the interesting comparison results for different methods.

Definition 5.1 (Problem Definition) Given multiple hierarchies (DMOZ and Wikipedia), our goal is to develop models that can classify a master source database (DMOZ) with accuracy and efficiency (refer to Fig. 5.5). To improve the model accuracy, all methods (except STL) use an external source Wikipedia dataset in conjunction with DMOZ dataset. Using external source helps in learning the better model parameters for source classes, by increasing the number of positive examples, especially when the number of training examples for a given class is scarce.

Different Approaches for Model Learning

1. Multi-task learning (MTL)—In this method, related tasks for DMOZ and Wikipedia datasets are learned together as shown in Fig. 5.6. Task relationships

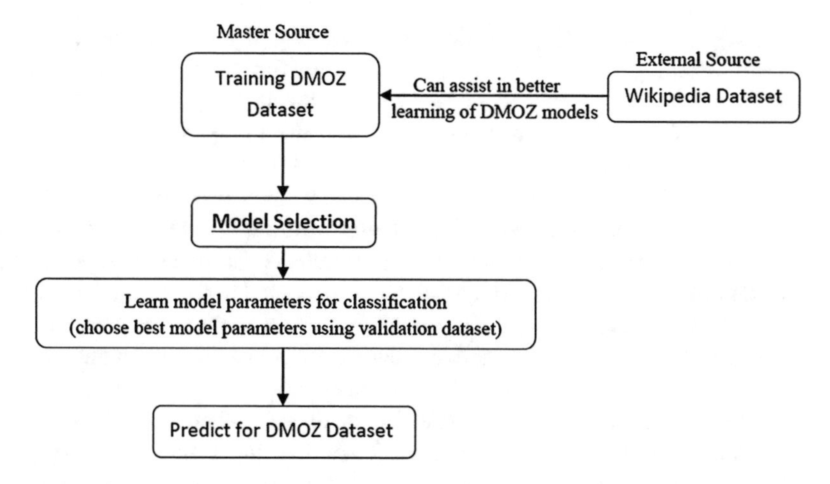

Fig. 5.5 Learn model to classify master source database DMOZ using external source database Wikipedia

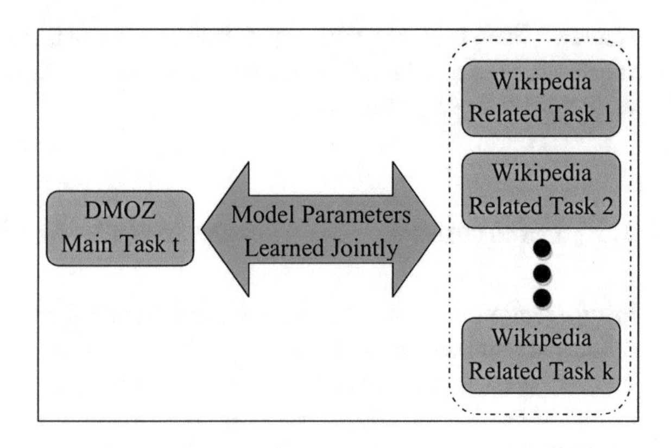

Fig. 5.6 Joint training in multi-task learning

across DMOZ and Wikipedia datasets are identified using the non-parametric, lazy nearest neighbor approach (kNN) [5]. k-related tasks are identified from Wikipedia dataset corresponding to each task in DMOZ dataset. Tanimoto similarity (Jaccard index) is used for computing the related tasks. Objective function for MTL is given by,

$$\min_{\mathbf{W}} \sum_{t=1}^{T} \sum_{i=1}^{N_t} \mathcal{L}\big(\mathbf{w}_t, \mathbf{x}_t(i), y_t(i)\big) + \lambda_1 ||\mathbf{w}_t||_2^2 + \lambda_2 \sum_{j=1}^{k} ||\mathbf{w}_{N_j(t)}||_2^2 + \lambda_3 \sum_{j=1}^{k} ||\mathbf{w}_t - \mathbf{w}_{N_j(t)}||_2^2 \quad (5.7)$$

where $\mathbf{w}_{N_j(t)}$ is the model parameters corresponding to $j-th$-related task.

2. Transfer learning (TL)—It is designed to learn the parameters of the main task (also know as parent task) based on the transferred parameters from the related task(s) (also known as children tasks) (refer to Fig. 5.7). Main intuition behind using TL is that the information contained in the children task can help in learning the better predictive models for the main task. When transferred parameters from the child task assist in better learning the predictive models of the parent task, then it is referred to as positive transfer. However, in some cases if related tasks are not found correctly, then TL may lead to the predictive models which are worse than the original predictive models without transfer, and this type of transfer is known as negative transfer. It has been shown in the work of Pan et al. [25] that TL improves the generalization performance of the predictive models provided the related tasks are similar to each other. The goal of TL is to learn the mapping function $f : X \rightarrow Y$ from N_t input-output pairs $\big\{\big(\mathbf{x}_t(1), y_t(1)\big), \big(\mathbf{x}_t(2), y_t(2)\big), \ldots, \big(\mathbf{x}_t(N), y_t(N)\big)\big\}$ in such a way so as to minimize the objective function given by,

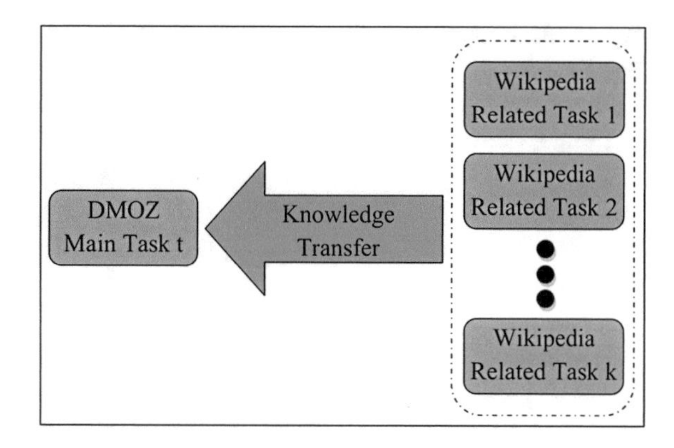

Fig. 5.7 Knowledge transfer in Transfer Learning

$$\min_{\mathbf{w}_t} \sum_{i=1}^{N_t} \mathscr{L}\big(\mathbf{w}_t, \mathbf{x}_t(i), y_t(i)\big) + \lambda_1 \big|\big|\mathbf{w}_t\big|\big|_2^2 + \lambda_2 \sum_{j=1}^{k} \big|\big|\mathbf{w}_t - \mathbf{w}^*_{N_j(t)}\big|\big|_2^2 \tag{5.8}$$

where $\mathbf{w}^*_{N_j(t)}$ is the learned model parameters that is transferred from the $j-th$ child (related) task.

TL differs from MTL in terms of parameter learning behavior. In MTL, all related task parameters are learned simultaneously, whereas in TL, related task parameters are learned first which is then transferred to the main task of interest. TL has also been referred to as asymmetric MTL because of its focus on one of the tasks, referred to as the parent (or main) task.

3. Semi-supervised learning (SSL)—It involves use of both labeled and unlabeled data for predicting the parameters of the model (Fig. 5.8). SSL falls in between unsupervised (no labeled training data) and supervised learning (completely labeled training data) [39]. SSL works on the principle that the more the training examples, the better the generality. However, the result of SSL is largely dependent on how accurately we group the unlabeled data with the labeled data. The more accurate the grouping of unlabeled data with labeled data, the better the result. For developing models for DMOZ classes using SSL method, kNN with Tanimoto similarity is used to find the groupings of unlabeled Wikipedia dataset with each of the DMOZ class. As such, objective function for learning $t-th$ task can be represented by,

$$\min_{\mathbf{w}_t} \sum_{i=1}^{N_t+\{N_j\}_{j=1}^{k}} \mathscr{L}\big(\mathbf{w}_t, \mathbf{x}_t(i), y_t(i)\big) + \lambda \big|\big|\mathbf{w}_t\big|\big|_2^2 \tag{5.9}$$

where N_j is the examples from the related tasks that are computed using kNN.

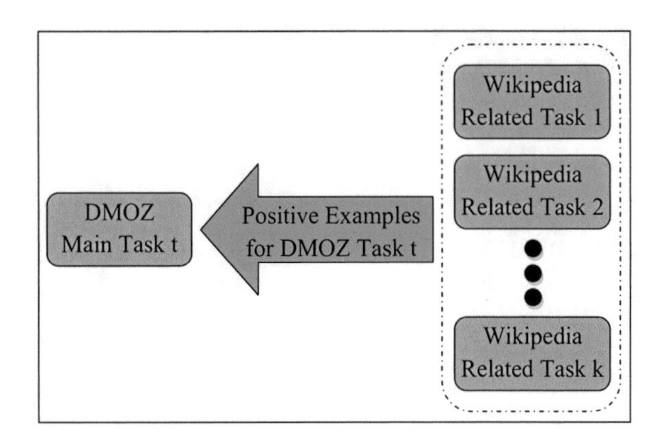

Fig. 5.8 Semi-supervised learning

5.5.1 Experimental Analysis

Dataset DMOZ and Wikipedia datasets used in the experiments are available from the ECML/PKDD 2012 discovery challenge on LSHTC (Track 2 challenge website).[1] The challenge is closed for new submission and the labels of the test set are not publicly available. Therefore, original training set is split into three parts (train, validation, test) in the ratio 3:1:1 for performance analysis. To assess the performance of different methods with respect to the class size (number of training examples per class), training data is further categorized into low distribution (LD), with 25 examples per class, and high distribution (HD), with 250 examples per class. This resulted in DMOZ dataset having 75 classes within LD and 53 classes within HD.

Results Figures 5.9, 5.10, 5.11, and 5.12 give the performance comparison of different methods on low distribution and high distribution DMOZ datasets. The following observations are worth noting from the results.

- STL vs SSL vs TL vs MTL: For LD dataset MTL method significantly outperforms other methods with improvement up to 15% in $\mu F1$ score. This is because joint learning of classes with scarce examples benefits from other related classes due to inductive transfer. For HD dataset, even though MTL has the best result, improvement is much smaller, i.e., $<$1% because with more training data, all model converges to similar performance.
- $k = 2$ vs $k = 3$ vs $k = 4$ vs $k = 5$ vs $k = 6$: In general, lower value of k gave better models compared to higher values of k. This is because as the value of k increases, similarity between the main tasks and the surrogate tasks decreases, which in turn affects the performance negatively.

Table 5.1 shows the average training time (in second) per class required to learn the models for the different LD and HD categories. The STL approach has the lowest

[1] http://lshtc.iit.demokritos.gr/LSHTC3_DATASETS.

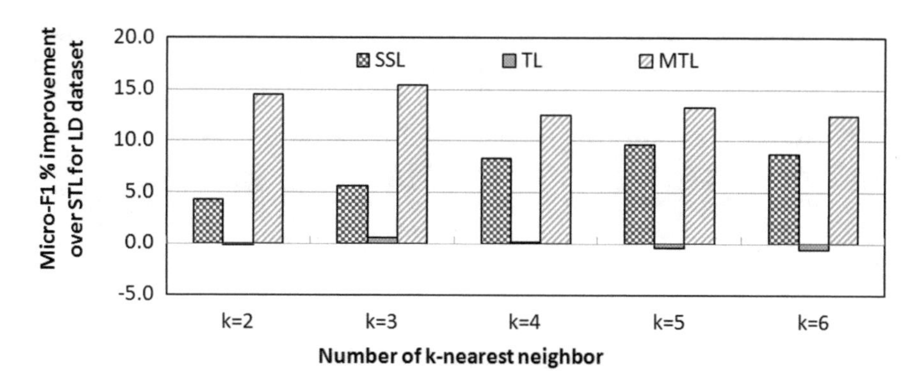

Fig. 5.9 Micro-F1 ($\mu F1$) comparison of STL against different methods for low distribution dataset

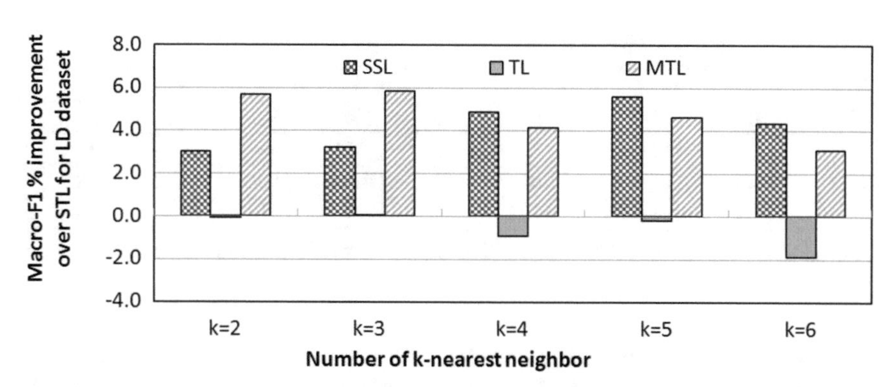

Fig. 5.10 Macro-F1 (MF1) comparison of STL against different methods for low distribution dataset

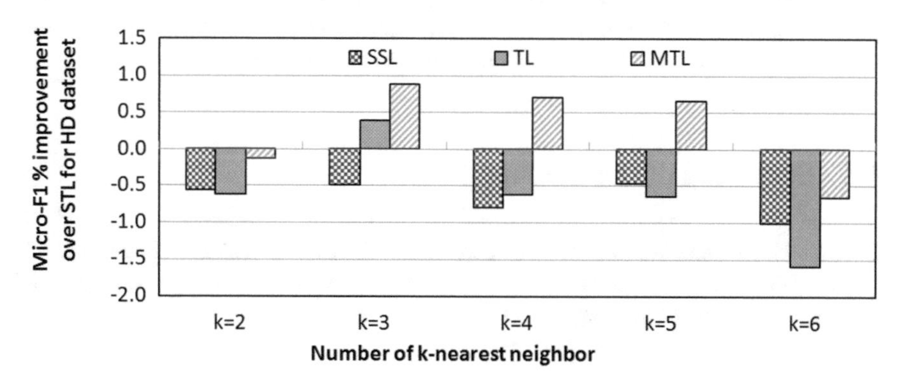

Fig. 5.11 Micro-F1 ($\mu F1$) comparison of STL against different methods for high distribution dataset

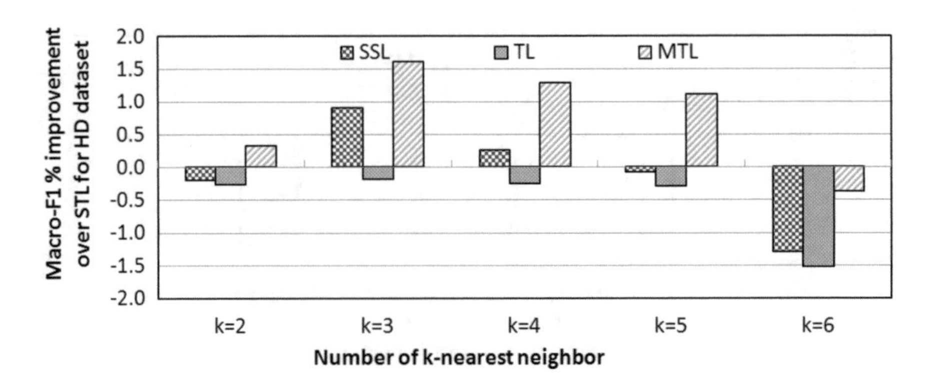

Fig. 5.12 Macro-F1 (MF1) comparison of STL against different methods for high distribution dataset

Table 5.1 Total training time (in second) for DMOZ dataset

STL		# of	SSL		TL		MTL	
LD	HD	*k*NN	LD	HD	LD	HD	LD	HD
2.72	44.7	(k = 2)	4.58	62.4	4.54	48.4	9.84	56.8
		(k = 4)	5.57	62.7	6.40	50.3	15.6	78.7
		(k = 6)	6.48	64.3	7.98	54.7	18.8	82.8

STL learns model for each DMOZ class without using Wikipedia dataset; hence it is independent of *k*NN value

training times because there is no overhead of incorporating additional constraints that is involved. SSL models take more time than the corresponding STL models because of the increased number of training examples. For TL models as well, runtime increases because it requires learning the models for the neighbors. Finally, MTL method takes the longest time, since it requires the joint learning of the model parameters that are updated for each class and related neighbors.

5.6 Summary of the Chapter

This chapter covers various ways to perform MTL for HC problem. Experimental results show that MTL is useful for improving the generalization performance of learned models but at the cost of increased runtime complexity.

References

1. Argyriou, A., Evgeniou, T., Pontil, M.: Multi-task feature learning. NIPS p. 19:41 (2007)
2. Argyriou, A., Evgeniou, T., Pontil, M.: Convex multi-task feature learning. Machine Learning **73(3)**, 243–272 (2008)

3. Baxter, J.: A model of inductive bias learning. JAIR **12**, 149–198 (2000)
4. Ben-David, S., Schuller, R.: Exploiting task relatedness for multiple task learning. Learning Theory and Kernel Machines pp. 567–580 (2003)
5. Bhatia, N., et al.: Survey of nearest neighbor techniques. arXiv preprint arXiv:1007.0085 (2010)
6. Bonilla, E., Chai, K., Williams, C.: Multi-task gaussian process prediction. NIPS (20(October), 2008)
7. Caruana, R.: Multitask learning. Machine Learning **28(1)**, 41–75 (1997)
8. Charuvaka, A., Rangwala, H.: Multi-task learning for classifying proteins using dual hierarchies. In: Proceedings of the 12th International Conference on Data Mining, pp. 834–839 (2012)
9. Charuvaka, A., Rangwala, H.: Multi-task learning for classifying proteins with dual hierarchies. pp. 834–839 (2012)
10. Collobert, R., Weston, J.: A unified architecture for natural language processing: Deep neural networks with multitask learning. In: Proceedings of the 25th international conference on Machine learning, pp. 160–167 (2008)
11. Deng, L., Hinton, G., Kingsbury, B.: New types of deep neural network learning for speech recognition and related applications: An overview. In: Acoustics, Speech and Signal Processing (ICASSP), 2013 IEEE International Conference on, pp. 8599–8603 (2013)
12. Evgeniou, T., Micchelli, C., Pontil, M.: Learning multiple tasks with kernel methods. JMLR **6(1)**, 615–637 (2005)
13. Evgeniou, T., Pontil, M.: Regularized multitask learning. KDD pp. 109–117 (2004)
14. Ghosn, J., Bengio, Y.: Multi-task learning for stock selection. Advances in Neural Information Processing Systems pp. 946–952 (1997)
15. Girshick, R.: Fast r-cnn. In: Proceedings of the IEEE international conference on computer vision, pp. 1440–1448 (2015)
16. Jacob, L., Bach, F., Vert, J.: Clustered multi-task learning: A convex formulation. NIPS (2008)
17. Jebara, T.: Multi-task feature and kernel selection for svms. ICML p. 55 (2004)
18. Jebara, T.: Multitask sparsity via maximum entropy discrimination. JMLR pp. 75–110 (2011)
19. Kato, T., Kashima, H., Sugiyama, M., Asai, K.: Multi-task learning via conic programming. NIPS pp. 737–744 (2008)
20. Leslie, C., Eskin, E., Noble, W.S.: The spectrum kernel: A string kernel for svm protein classification. In: Biocomputing 2002, pp. 564–575. World Scientific (2001)
21. Liu, J., Ji, S., Ye, J.: Multi-task feature learning via efficient l 2, 1-norm minimization. UIA pp. 339–348 (2009)
22. Murzin, A.G., Brenner, S.E., Hubbard, T., Chothia, C.: Scop: a structural classification of proteins database for the investigation of sequences and structures. Journal of molecular biology **247**(4), 536–540 (1995)
23. Obozinski, G., Taskar, B., Jordan, M.: Multi-task feature selection. ICML (2006)
24. Orengo, C.A., Michie, A., Jones, S., Jones, D.T., Swindells, M., Thornton, J.M.: Cath–a hierarchic classification of protein domain structures. Structure **5**(8), 1093–1109 (1997)
25. Pan, S., Yang, Q.: A survey on transfer learning. IEEE Transactions on Knowledge and Data Engineering **22(10)**, 1345–1359 (2010)
26. Pan, S.J., Yang, Q., et al.: A survey on transfer learning. IEEE Transactions on knowledge and data engineering pp. 1345–1359 (2010)
27. Pong, T.K., Tseng, P., Ji, S., Ye, J.: Trace norm regularization: Reformulations, algorithms, and multi-task learning. SIAM Journal on Optimization **20**(6), 3465–3489 (2010)
28. Ramsundar, B., Kearnes, S., Riley, P., Webster, D., Konerding, D., Pande, V.: Massively multitask networks for drug discovery. arXiv preprint arXiv:1502.02072 (2015)
29. Thrun, S.: Is learning the n-th thing any easier than learning the first? NIPS pp. 640–646 (1996)
30. Thrun, S., Sullivan, J.O.: Clustering learning tasks and the selective cross-task transfer of knowledge. Learning to learn pp. 181–209 (1998)
31. Tibshirani, R.: Regression shrinkage and selection via the lasso. Journal of the Royal Statistical Society. Series B (Methodological) pp. 267–288 (1996)

32. Wang, X., Zhang, C., Zhang, Z.: Boosted multi-task learning for face verification with applications to web image and video search. IEEE conference on Computer Vision and Pattern Recognition pp. 142–149 (2009)
33. Widmer, C., Leiva, J., Altun, Y., Rätsch, G.: Leveraging sequence classification by taxonomy-based multitask learning. 14th Annual International Conference, RECOMB, Lisbon, Portugal pp. 522–534 (April 25–28, 2010)
34. Yu, S., Yu, K., Tresp, V., Kriegel, H.: Collaborative ordinal regression. In Proceedings of the 23rd international conference on Machine learning pp. 1089–1096 (2006)
35. Zhang, Y., Yeung, D.: A convex formulation for learning task relationships in multi-task learning. In Proceedings of the Twenty-fourth Conference on Uncertainty in AI (UAI) (2010)
36. Zhou, J., Chen, J., Ye, J.: Clustered multi-task learning via alternating structure optimization. NIPS (2011)
37. Zhou, J., Chen, J., Ye, J.: MALSAR: Multi-tAsk Learning via StructurAl Regularization. Arizona State University (2011). URL http://www.public.asu.edu/~jye02/Software/MALSAR
38. Zhou, J., Lei, Y., Liu, J., Ye, J.: A multi-task learning formulation for predicting disease progression. In Proceedings of the 17th ACM SIGKDD international conference on Knowledge discovery and data mining pp. 814–822 (2011)
39. Zhu, X.: Semi-supervised learning literature survey. world **10**, 10 (2005)

Chapter 6
Conclusions and Future Research Directions

There has been tremendous work in the field of HC and it's not possible to cover everything. In this book, we discussed about various problems that are associated with LSHC along with different approaches to solve it. Specifically, two of the chapters are devoted to discuss in detail about hierarchical inconsistencies and feature selection problems. We also discussed about the MTL learning paradigm that leverage multiple hierarchies classifying the similar types of data. Overall, our main intention in this book was to provide comprehensive overview about the LSHC literature that we believe would be helpful for the readers with intermediate expertise in data mining having a background in classification (supervised learning). We have put an extra effort to present the contents in as simplistic manner as possible, providing examples and figures whenever we felt it would be helpful in understanding the concepts. We hope that readers enjoyed reading this book and gained substantial knowledge about advancements in LSHC field. Finally, we would like to encourage the readers to read the paper [1] for gaining better insights about when to use flat and hierarchical methods for newly created hierarchical datasets.

6.1 Future Research Directions

For interested readers, this section provides useful guidance about various future research directions for extending LSHC research.

6.1.1 Extreme Classification

Overtime, the number of labels (categories) keeps on increasing. Extreme classification is the problem of dealing with extremely large label spaces. To motivate

A. Naik, H. Rangwala, *Large Scale Hierarchical Classification: State of the Art*, SpringerBriefs in Computer Science, https://doi.org/10.1007/978-3-030-01620-3_6

consider the example of social media such as Twitter where new hashtags are being created by the users frequently. Obviously, when new hashtag appears, they do not have enough instances to train generalized models for classifying future tweets. In such cases it would be beneficial to fetch instances from other related hashtags. Problem that would be interesting to solve is how to determine the related hashtags from such an extremely large space of hashtags. Many works in this direction have been proposed in the literature [2, 4, 7, 10, 14, 18]. Similar approach can be extended to the HC problem where labels (tags) can be organized into the hierarchy, and mapping of unlabeled tags can be done easily by recursively selecting the best set of tags in the hierarchy.

Extreme classification problem: given instance, find its most relevant subset of labels from extremely large label space.

At another level, given streaming data, the following situation can arise which will lead to interesting learning problems.

1. Assume we are given streaming data with tweets. Before a hashtag gets popular, users may choose not to use a particular one of interest at all or make up their own. In this setting, can we reassign previous tweets before the hashtag originated to the particular hashtag using classification?
2. Given the limited data and treating hashtag assignments as an extreme classification problem, can we use the data from step 1 to make predictions on future tweets? Can we know that a tweet generated needs a new hashtag that does not exist in the previous pool, i.e., orphan prediction, in the classification literature?

6.1.2 *Partial Flattening of Inconsistent Nodes*

Removing inconsistent nodes from the original (expert defined) hierarchy is beneficial for improving the HC performance [1, 16, 19]. However, flattening all children of the inconsistent nodes may not be an optimal choice for classification. It is quiet possible that some of the children of inconsistent nodes are actually benefiting from these nodes by leveraging structural information (especially rare categories). However, due to overall optimization objective value, the node is identified as inconsistent, and all its children are flattened. To overcome this drawback, there is a scope to perform more regressive analysis with partial flattening of inconsistent nodes where only the subset of children are flattened as shown in Fig. 6.1.

Validation dataset can be used to identify the subset of children that performs comparatively better with the inconsistent nodes' presence.

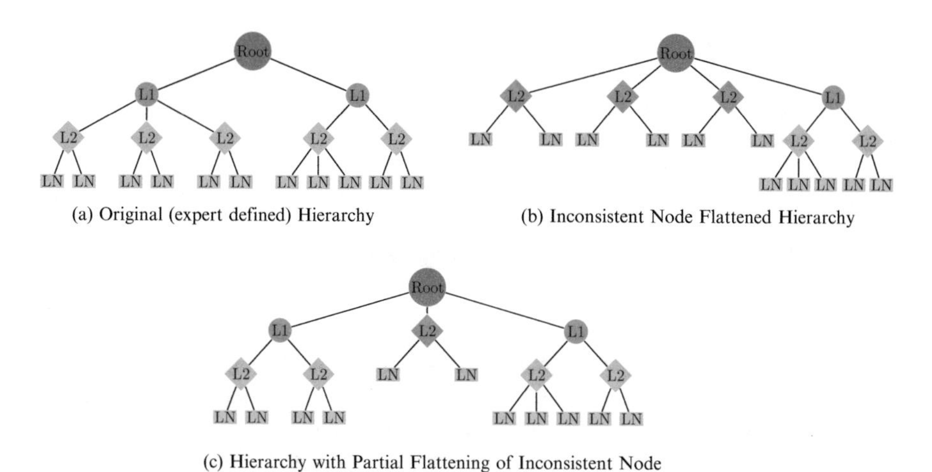

Fig. 6.1 Different hierarchical structures (**b**)–(**c**) obtained by flattening original (expert defined) hierarchy (**a**)

6.1.3 Multi-Linear Models

As stated in earlier chapters, TD methods are effective for dealing with LSHC. However, the performance of TD is poor due to error propagation, i.e., errors made at the higher levels in the hierarchy cannot be corrected at the lower levels. To overcome this problem, various methods have been proposed that modify hierarchy [1, 13] or combine multiple predictions [3, 6] for improving the classification performance. Still, the margin of errors at higher levels in the hierarchy is more compared to lower levels [16] because at higher levels, each of the discriminative node consists of the multiple subcategories which may not be easily separable with the linear classifiers. Alternatively, nonlinear classifiers [5] can be used to train models at higher levels, but they are computationally expensive which makes them unsuitable for large-scale problem.

An ideal classifier must possess the classifying properties of nonlinear classifiers while being computationally efficient like linear classifiers. Recently, multi-linear methods have been proposed by Huang and Lin [9] which address this issue for binary classification problem. Multi-linear models take advantage of both linear and nonlinear methods. While multi-linear models are more accurate than linear models, it is also computationally efficient than nonlinear models. One of the logical research extensions of LSHC work could be to explore multi-linear models for multi-class HC problem.

6.1.4 Detecting New Categories

Data distribution changes over time, and we need to evolve the hierarchy based on the new unseen data. When a new category is emerging, it is very likely that instances from that category cannot be confidently predicted into any of the subcategories at lower level but stay at the higher-level categories. Detecting such emerging topics (orphan node prediction problem) is crucial for maintaining the effectiveness of HC.

6.1.5 Feature Representation Using Deep Learning

In the last few years, deep learning have emerged as a powerful machine learning tool and has proven to be effective for object recognition in computer vision and speech recognition problems [12]. In order to achieve superior HC performance, deep learning can be exploited for better hierarchical feature representation. Recently, there has been some work around deep learning to address LSHC problem [11, 17]. However, there are plenty of scopes for further improvements.

6.1.6 Large-Scale Multi-task Learning

One of the major problems with MTL is that it is computationally expensive due to joint learning and cannot be scaled for large-scale problems. For reducing the runtime performance of MTL-based methods, feature selection could be an effective tool [8, 15, 20]. Large-scale problems have high-dimensional features which increase the runtime between optimization iterations resulting in longer runtime. Feature selection would be helpful to reduce the dimensions of instances, thereby reducing the time taken to complete each iteration and hence improving the runtime performance. It would also result in improved accuracy (by removing the effect of irrelevant features) and lesser memory to store the model parameters for large-scale datasets.

References

1. Babbar, R., Partalas, I., Gaussier, E., Amini, M.R.: On flat versus hierarchical classification in large-scale taxonomies. In: Advances in Neural Information Processing Systems, pp. 1824–1832 (2013)
2. Belanger, D., McCallum, A.: Structured prediction energy networks. arXiv preprint arXiv:1511.06350 (2015)
3. Bennett, P.N., Nguyen, N.: Refined experts: improving classification in large taxonomies. In: Proceedings of the 32nd international ACM SIGIR conference on Research and development in information retrieval, pp. 11–18 (2009)
4. Bhatia, K., Jain, H., Kar, P., Varma, M., Jain, P.: Sparse local embeddings for extreme multi-label classification. In: Advances in Neural Information Processing Systems, pp. 730–738 (2015)

5. Burges, C.J.: A tutorial on support vector machines for pattern recognition. Data Mining and Knowledge Discovery **2**(2), 121–167 (1998)
6. Cesa-Bianchi, N., Gentile, C., Zaniboni, L.: Hierarchical classification: combining bayes with svm. In: Proceedings of the 23rd International Conference on Machine Learning (ICML), pp. 177–184 (2006)
7. Dekel, O., Shamir, O.: Multiclass-multilabel classification with more classes than examples. In: AISTATS, pp. 137–144 (2010)
8. Heisele, B., Serre, T., Prentice, S., Poggio, T.: Hierarchical classification and feature reduction for fast face detection with support vector machines. Pattern Recognition **36**(9), 2007–2017 (2003)
9. Huang, H.Y., Lin, C.J.: Linear and kernel classification: When to use which? In: SIAM International Conference on Data Mining (SDM) (2016)
10. Jain, H., Prabhu, Y., Varma, M.: Extreme multi-label loss functions for recommendation, tagging, ranking & other missing label applications. In: Proceedings of the 22nd ACM SIGKDD International Conference on Knowledge Discovery and Data Mining, pp. 935–944 (2016)
11. Kowsari, K., Brown, D.E., Heidarysafa, M., Meimandi, K.J., Gerber, M.S., Barnes, L.E.: Hdltex: Hierarchical deep learning for text classification. In: Machine Learning and Applications (ICMLA), 2017 16th IEEE International Conference on, pp. 364–371 (2017)
12. LeCun, Y., Bengio, Y., Hinton, G.: Deep learning. nature **521**(7553), 436 (2015)
13. Malik, H.: Improving hierarchical svms by hierarchy flattening and lazy classification. In: Large-Scale Hierarchical Classification Workshop of ECIR (2010)
14. Mineiro, P., Karampatziakis, N.: A hierarchical spectral method for extreme classification. http://arxiv.org/abs/1511.03260 (2016)
15. Naik, A., Rangwala, H.: Embedding feature selection for large-scale hierarchical classification. In: Proceedings of the IEEE International Conference on Big Data, pp. 1212–1221 (2016)
16. Naik, A., Rangwala, H.: Inconsistent node flattening for improving top-down hierarchical classification. In: IEEE International Conference on Data Science and Advanced Analytics (DSAA), pp. 379–388 (2016)
17. Peng, H., Li, J., He, Y., Liu, Y., Bao, M., Wang, L., Song, Y., Yang, Q.: Large-scale hierarchical text classification with recursively regularized deep graph-cnn. In: Proceedings of the 2018 World Wide Web Conference on World Wide Web, pp. 1063–1072 (2018)
18. Prabhu, Y., Varma, M.: Fastxml: A fast, accurate and stable tree-classifier for extreme multi-label learning. In: Proceedings of the 20th ACM SIGKDD International Conference on Knowledge Discovery and Data mining, pp. 263–272 (2014)
19. Wang, X.L., Lu, B.L.: Flatten hierarchies for large-scale hierarchical text categorization. In: Proceedings of the fifth International Conference on Digital Information Management (ICDIM), pp. 139–144 (2010)
20. Zhou, D., Xiao, L., Wu, M.: Hierarchical classification via orthogonal transfer. In: Proceedings of the 28th International Conference on Machine Learning (ICML), pp. 801–808 (2011)

Printed in the United States
By Bookmasters